HARCOURT BRACE & COMPANY
1919-
1994
SEVENTY-FIVE YEARS

My Tears Spoiled My Aim

BOOKS BY JOHN SHELTON REED

WHISTLING DIXIE: Dispatches from the South

SOUTHERN FOLK: Plain and Fancy

SOUTHERNERS: The Social Psychology of Sectionalism

ONE SOUTH: An Ethnic Approach to Regional Culture

REGIONALISM AND THE SOUTH: Selected Papers of Rupert Vance
(editor, with Daniel Joseph Singal)

PERSPECTIVES ON THE AMERICAN SOUTH
(editor, with Merle Black)

THE ENDURING SOUTH: Subcultural Persistence in Mass Society

My Tears Spoiled My Aim

and Other Reflections on Southern Culture

John Shelton Reed

A Harvest Book
Harcourt Brace & Company
San Diego New York London

Library of Congress Cataloging-in-Publication Data
Reed, John Shelton.
My tears spoiled my aim, and other reflections on Southern culture/John Shelton Reed.
p. cm. — (A Harvest book)
ISBN 0-15-600006-7
1. Southern States — Civilization — 20th century. 2. Southern
States — Social conditions. I. Title.
[F216.2.R42 1994] 94-6750
978 — dc20

Printed in the United States of America
First Harvest edition 1994
A B C D E

To the University of North Carolina
on its bicentennial,
with gratitude

Contents

Acknowledgments

■ Nearly all the chapters in this book began as lectures, either to a variety of academic and popular audiences or to my undergraduate class on the South. They have been improved by the questions and suggestions of a great many listeners, whom I thank for their interest and courtesy. Those origins also explain, if not excuse, a certain casualness when it comes to citation, but I have tried to provide enough information in the text so that readers can use the bibliography to follow up on any points of interest. I am grateful to the many colleges, universities, and other organizations who have, from time to time, forced me to organize my thoughts and observations on some subject by asking me to speak about it.

Several chapters were written while I was a fellow at the National Humanities Center, and editing them for publication was one of the activities supported by a fellowship at the Center for Advanced Study in the Behavioral Sciences. I thank the staffs of these two splendid retreats for their hospitality and assistance. Those fellowships were funded by National Endowment for the Humanities grant FC-2038-81 and National Science Foundation grant BNS-8700864, respectively, and some of the data reported in "Life and Leisure in the New South" were produced under National Endowment for the Arts grant 02-4050-005—which makes what is called in hockey a "hat trick." I appreciate the support, however involuntary, of the long-suffering taxpayers of the United States.

Helpful early readings of the manuscript were provided by two distinguished Southern historians, Wayne Flynt of Auburn University and James Cobb of the University of Tennessee, as well as by my editor, Beverly Jarrett, director of the University of Missouri Press. All three recognize that friendship doesn't require uncritical enthusiasm, and this book is better for their suggestions. Dale Volberg

Reed has also been a discerning reader, as always, and I thank her for that, among many other things.

For more than twenty years the staff of the Institute for Research in Social Science at the University of North Carolina at Chapel Hill has been unfailingly and invaluably helpful in my research, and their assistance is reflected in almost every chapter. Particular help with the preparation of this book has come from Mike Crane and Alecia Holland, and it is a pleasure to thank them here. It has also been a pleasure to work again with the staff of the University of Missouri Press, especially Jane Lago, a tactful and superbly competent copy editor.

I thank Athan Manuel and Charles Wilson for their permission to reprint our co-authored essay here, and I also thank the publishers of those chapters that have already appeared in print, as follows: The University of North Carolina Press for "The South: What Is It? *Where* Is It?" from Paul Escott and David Goldfield, eds., *The South for New Southerners* (1991). Harvard University Press for "Southerners as an American Ethnic Group," originally published as "Southerners," in S. Thernstrom, ed., *The Harvard Encyclopedia of American Ethnic Groups* (1980). *Southern Humanities Review* for "The South's Mid-Life Crisis," from volume 25 (Spring 1991): 125–35. The University of Alabama Press for "New South or No South?: Southern Culture in 2036," from J. Himes, ed., *The South Moves into Its Future* (1991). The Southern Growth Policies Board for "Southern Public Opinion on the South's Quality of Life," from Commission on the Future of the South, *Education, Environment and Culture: The Quality of Life in the South* (1986). *North Carolina Historical Review* for "Life and Leisure in the New South," from volume 60 (April 1983): 172–82. Gordon & Breach, Science Publishers, for "*Playboy*'s Southern Exposure" [with Athan Manuel and Charles R. Wilson], from James Cobb and Charles R. Wilson, eds., *Perspectives on the American South: An Annual Review of Society, Politics and Culture,* vol. 4 (1987). Southern Progress Corporation for "In Search of the Elusive Southerner," from *Southern Living,* June 1990, pp. 92–94. Parts of "Thoughts on the Southern Diaspora" and "Refugees and Returnees" have appeared in my column, "Letter from the Lower Right," in *Chronicles: A Magazine of American Culture.*

I would like to be thanking a number of music publishers for permission to quote copyrighted song lyrics in this book's title-essay, except that such permission proved so expensive or difficult to procure that I have rewritten the essay instead. I can understand why Hank Williams, Jr.'s publishers wanted $800 for my paltry "use of Hank's personal creativity" (young Hank recently lost half his daddy's estate in a lawsuit and probably needs the money), but I do not understand publishers who do not answer their mail or return telephone calls. Anyway, I am truly grateful for permission to quote from the following:

My Tears
Spoiled My Aim

Introduction

> That a change is now in course all over the South is plain; and it is as plain that the South changing must be the South still.
>
> **—Stark Young, in *I'll Take My Stand* (1930)**

■ It has been twenty-odd years now since some of us began to take seriously the similarities between Southerners and members of groups more often regarded as "ethnic." That approach to the study of the South had been suggested decades earlier, but it was only in the context of the so-called ethnic revival of the late 1960s that I, among many others, really started to ask the same questions about Southerners that had long been asked about American immigrant groups. One result was my book *One South: An Ethnic Approach to Regional Culture,* a collection of essays and addresses treating contemporary Southern culture and identity from the peculiar viewpoint of a sociologist trained in public opinion research and interested in the general topic of ethnicity.

Naturally I was delighted when John Boles remarked in a review of that book that the ethnic analogy had proved "a fruitful way of analyzing the persistence of Southern identity," in fact "the most satisfying modern interpretation of the South." Not everyone agrees, I suppose, but at least that analogy was an idea whose time had come, occurring as it did to many of us more or less independently—to Lewis Killian, Andrew Greeley, Norval Glenn, and my Chapel Hill colleague George Tindall, among others. I differ from those other scholars mostly in that I've done little else since but explore the implications of that one apparently good idea.

So this volume is almost a sort of appendix to *One South*—so much so that I thought about calling it *Two South,* to be followed (readers of Dr. Seuss will have already guessed) by *Red South* and *Blue South.* But when cooler heads prevailed, I proposed a title taken from my epigraph above: I thought that *The South Changing, The South Still* nicely summarized the juxtaposition of transformative change and continuing identity (a combination that could be called paradoxical if it were not common to all living organisms). When some readers of the manuscript objected to that title, too—for one at least it evoked images of the illicit distillation of spirits—we settled on the present one, which at least informs readers up front that they are holding a collection rather than a monograph.

Several of the essays in this collection explore aspects of the ethnic analogy, looking directly at questions of identity, cultural distinctiveness, boundary maintenance, and the like. Others merely follow from the implication that what is important is less the *region* than the *regional group* that it defines, and that defines it. The first chapter, for instance, argues in effect that "the South" is an increasingly troublesome concept—unless it is defined as "where Southerners are found." Even the chapter's origins reflect an interest in regional groups: "The South: What Is It? *Where* Is It?" began as part of a lecture series that toured a half-dozen cities of the Carolinas, called "The South for Non-Southerners." (When the lectures were published as a book, the publishers gave it what they took to be a more hospitable title, *The South for* New *Southerners.*)

The second chapter, "Southerners as an American Ethnic Group," is probably the most explicit available statement of the ethnic analogy. Its history, too, may be instructive. Asked by the editors of *The Harvard Encyclopedia of American Ethnic Groups* to write on American regional groups, I soon discovered that few scholars have had much to say—or at least that I could not find much to say—about regional groups other than Southerners. After some negotiation, the editors agreed to a reduced assignment, and this was the result. (By the way, I am immoderately pleased to have been given a photograph of President Carter examining a presentation copy of the encyclopedia, open to this entry.)

"The South's Mid-Life Crisis," the next chapter, suggests some of the limitations of the ethnic analogy, examining what I take to be the changing basis of Southern regional identity, as the South's unique history becomes less and less consequential. This essay began its life as one of three Lamar Lectures at Georgia Wesleyan College in 1984, although I have given it in one form or another in a great many other places since.

The next four chapters examine either the existence or the implications of Southern cultural distinctiveness. "New South or No South?: Southern Culture in 2036," for instance, asks whether the South is becoming more like the rest of the country. Concluding, perhaps disappointingly, that the question can only be answered "yes and no," I try to explain why some differences are going away and others are not, to point out some ways in which the rest of the United States is starting to look like the South, and to suggest why humility is in order when it comes to making predictions. This, too, was originally a Lamar Lecture at Georgia Wesleyan, although after a good many revisions it wound up in print as part of a symposium to mark the fiftieth anniversary of the Southern Sociological Society.

"Preserving the South's Quality of Life" examines Southerners' considerable affection for the South and looks at why some aspects of Southern culture may make it difficult to preserve what Southerners say they like about their region. This chapter was originally a report prepared for the 1986 Commission on the Future of the South, convened by the Southern Growth Policies Board, and the quotations are from my *Southerners: The Social Psychology of Sectionalism.*

"Life and Leisure in the New South" examines a less troubling aspect of regional culture, the use of leisure time. (Historians will recognize the allusion to Ulrich Phillips's classic *Life and Labor in the Old South.*) This essay draws in part on research Peter Marsden and I did for the National Endowment for the Arts, and it started out as an address to the North Carolina Literary and Historical Society.

"My Tears Spoiled My Aim: Violence in Country Music" was written originally for a collection of scholarly essays on country music but was rejected when a reviewer complained that it was not "theoretical" enough. I resisted the temptation to inject the required theory: what happens when literary theory meets country music is not a pretty sight. This chapter is published here for the first time, in the belief that, however unsophisticated, it graphically illustrates some familiar points about violence in Southern culture.

Like "real" ethnic groups, Southerners have developed a taxonomy of social types, shorthand labels for the different kinds of group members one is likely to encounter. "*Playboy*'s Southern

Exposure" was written when the Lemon Lecture at Clemson University gave me an opportunity to try out some ideas on this subject, later elaborated in *Southern Folk, Plain and Fancy.* It incorporates data from a term paper by one of my undergraduate students, Athan Manuel, and some observations from Charles R. Wilson.

All ethnic groups have to address the phenomenon of boundary-crossing when members leave the group and nonmembers join it, or aspire to. In the case of a regional group, the boundaries are literal—geographical. Asked to give two Franklin Lectures at Auburn University in 1990, I looked for a topic that divided neatly into halves, and came up with migration: to the South, and from it. "Thoughts on the Southern Diaspora" and "Refugees and Returnees" look at some of the effects of interregional migration, at mechanisms operating to produce and to retard assimilation, and at strategies for coping with the unassimilated.

Finally, "In Search of the Elusive Southerner" is a coy little piece written for the twenty-fifth anniversary issue of *Southern Living* magazine. The conversation it reports really did take place; it is touched up only a bit for publication.

The South

What Is It? *Where* Is It?

■ So you've moved, or been moved, to the South. Or maybe you're thinking about it. You're wondering: What is this place? What's different about it? *Is* it different, anymore?

Good questions. Old ones, too. People have been asking them for decades. Some of us even make our livings by asking them, but we still don't agree about the answers. Let's look at what might seem to be a simpler question: Where is the South?

That's easy enough, isn't it? People more or less agree about which parts of the United States are in the South and which aren't. If I gave you a list of states and asked which are "Southern," all in all, chances are you'd agree with some of my students, whose answers are summarized in Figure 1. I don't share their hesitation about Arkansas, and I think too many were ready to put Missouri in the South, but there's not a lot to argue with here.

That tells us something. It tells us that the South is, to begin with, a concept—and a shared one. It's an idea that people can talk about, think about, use to orient themselves and each other. People know whether they're in it or not. As a geographer would put it, the South is a "vernacular" region.

Stop and think about that. Why should that be? Why can I write "South" with some assurance that you'll know I mean Richmond and don't mean Phoenix? What is it that the South's boundaries enclose?

Well, for starters, it's not news that the South has been an economically and demographically distinctive place—a poor, rural region with a biracial population, reflecting the historic dominance of the plantation system. One thing the South's boundaries have set

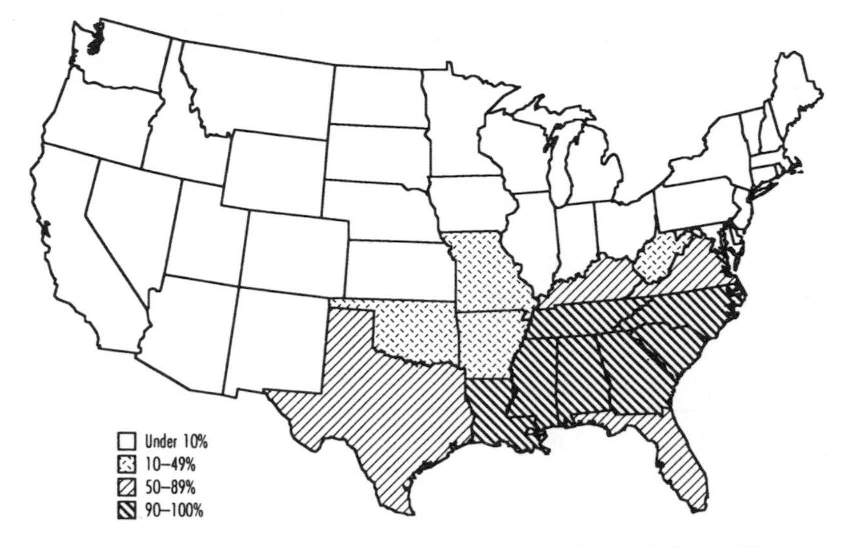

Figure 1. Percentage Who Say Each State Is Southern, "All in All"
Source: Sixty-eight students at the University of North Carolina at Chapel Hill.

off is a set of distinctive problems, growing out of that history. Those problems may be less and less obvious, but most are still with us to some extent, and we can still use them to locate the South.

But the South is more than just a collection of unfavorable statistics. It has also been home to several populations, black and white, whose intertwined cultures have set them off from other Americans as well as from each other. Some of us, in fact, have suggested that Southerners ought to be viewed as an American ethnic group, like Italian- or Polish-Americans. If we can use distinctive cultural attributes to find Southerners, then we can say that the South is where they are found.

Southerners are also like immigrant ethnic groups in that they have a sense of group identity based on their shared history and their cultural distinctiveness in the present. If we could get at it, one of the best ways to define the South would be with what Hamilton Horton calls the "Hell, yes!" line: where people begin to answer that way when asked if they're Southerners.

Finally, to the considerable extent that people do have a sense of the South's existence, its distinctiveness, and its boundaries, regional institutions have contributed. Southern businesses, Southern magazines, Southern voluntary associations, colleges, and universities—many such have at least aspired to serve the South as a whole. We can map the South by looking at where the influence of such enterprises extends.

All of these are plausible ways to go about answering the question of where the South is. For the most part, they give similar answers, which is reassuring. But it's where they differ (as they sometimes do) that they're most likely to tell us something about what the South has been, and is becoming. Nobody would exclude Mississippi from the South. But is Texas now a Southern state? Is Florida, anymore? How about West Virginia?

Allow me a homely simile. The South is like my favorite pair of blue jeans. It's shrunk some, faded a bit, got a few holes in it. There's always the possibility that it might split at the seams. It doesn't look much like it used to, but it's more comfortable, and there's probably a lot of wear left in it.

The Socioeconomic South

"Let us begin by discussing the weather," wrote U. B. Phillips in 1929. The weather, that distinguished Southern historian asserted, "has been the chief agency in making the South distinctive. It fostered the cultivation of the staple crops, which promoted the plantation system, which brought the importation of negroes, which not only gave rise to chattel slavery but created a lasting race problem. These led to controversy and regional rivalry for power, which . . . culminated in a stroke for independence." Phillips and the many who have shared his views see almost everything of interest about the South as emanating from this complex of plantation, black population, Civil War—thus, ultimately, from the weather.

You may have noticed that it's hot here in the summer, and humid. Some vegetable life loves that. Kudzu, for instance: that rampant, loopy vine needs long, moist summers, and gets them in the South. "Where kudzu grows" (Figure 2) isn't a bad definition of the South (and notice that it doesn't grow in southern Florida or West Texas).

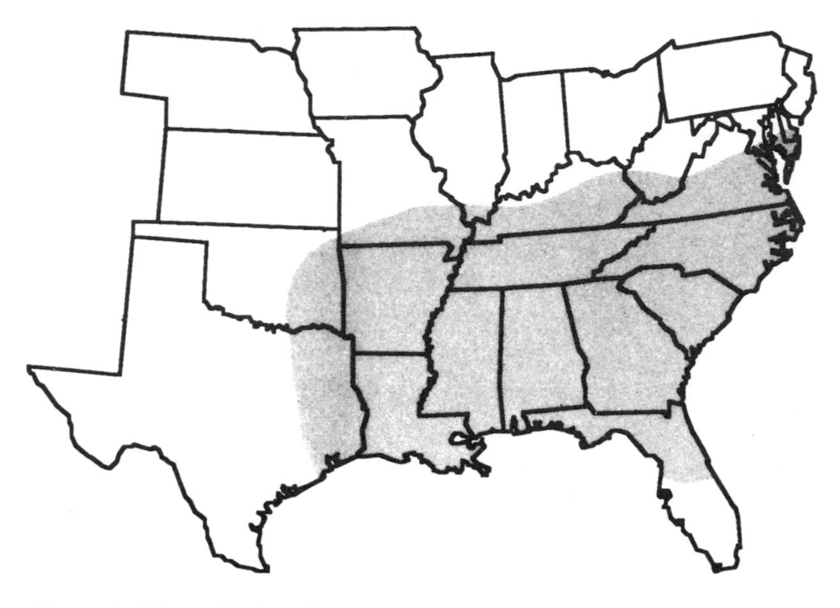

Figure 2. Where Kudzu Grows

Source: John J. Winberry and David M. Jones, "Rise and Decline of the 'Miracle Vine': Kudzu in the Southern Landscape," *Southeastern Geographer* 13 (November 1973): 62.

But another plant has been far more consequential for the South. That plant, of course, is cotton. Dixie *was* "the land of cotton," and Figure 3 shows that in the early years of this century Southerners grew cotton nearly everywhere they could grow it: everywhere with two hundred or more frost-free days, annual precipitation of twenty-three inches or more, and soil that wasn't sand.

Certainly cotton culture affected the racial makeup of the South and slowed the growth of Southern cities. Figure 4 shows what the region looked like, demographically, in 1920. Few cities interrupted the countryside. A band of rural counties with substantial black populations (solidly shaded on the map) traced the area of cotton cultivation and plantation agriculture, in a long arc from southeastern Virginia down and across to eastern Texas, with arms north and south along the Mississippi River.

This is the *Deep* South—what a geographer would call the "core

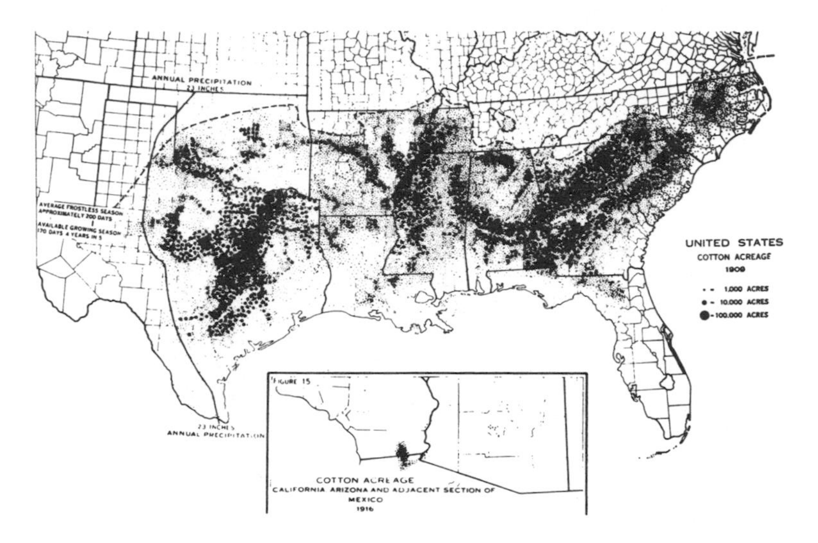

Figure 3. Acres of Cotton Cultivation, 1909

Source: U.S. Department of Agriculture, Bureau of Agricultural Economics.
Note: Map also shows gradients for precipitation and frost-free days.

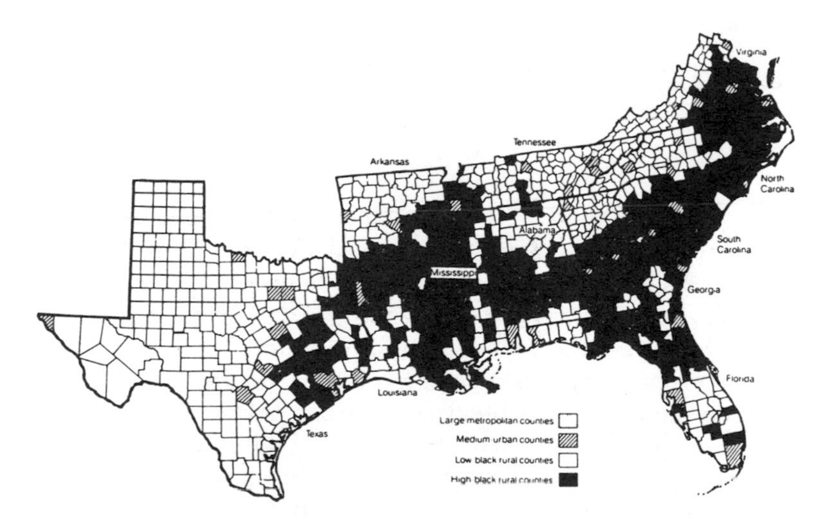

Figure 4. The Demography of the South, 1920

Source: Earl Black and Merle Black, *Politics and Society in the South* (Cambridge: Harvard University Press, 1987), 36.

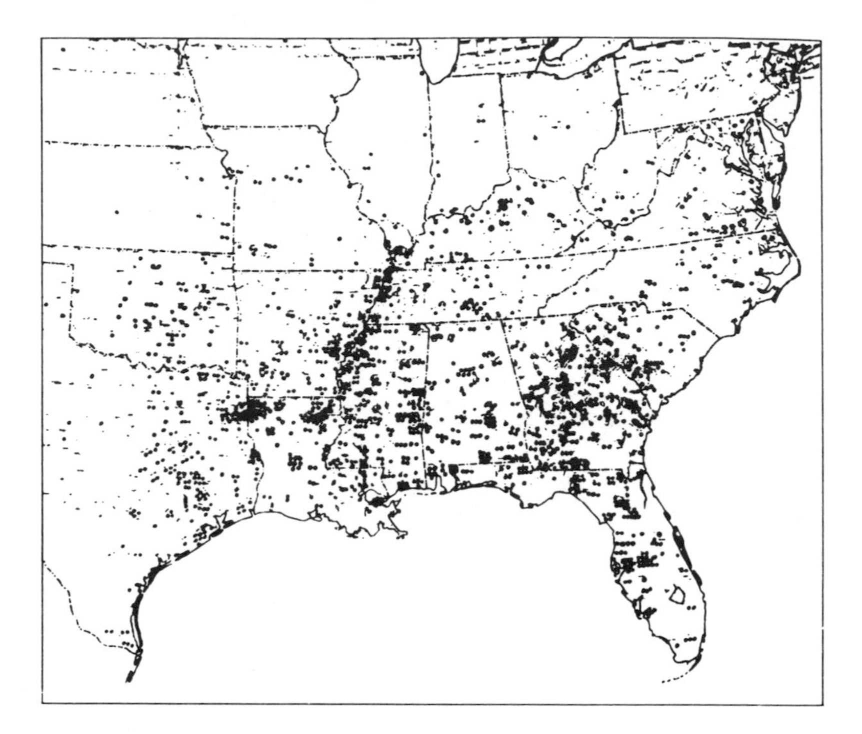

Figure 5. Lynchings, 1900–1930
Source: Southern Commission on the Study of Lynching.

area" of the region defined by its staple-crop economy. Here some Southern characteristics and phenomena were found in their purest, most concentrated form. Lynchings, for example (Figure 5). Or peculiar, single-issue politics (that issue, as a politician once put it, "spelled n-i-g-g-e-r"), reflected in support for third-party or unpopular major-party presidential candidates (Figure 6). For decades the Deep South shaped Southern culture and politics and, still more, shaped people's image of what the South was all about.

Two out of three Southerners are now urban folk, and most rural Southerners work in industry anyway, but the fossil remains of this old South can still be found as concentrations of poor, rural black Southerners (compare Figure 7, for 1980, to Figure 4). This population, together with poor, rural *white* Southerners in the Southern

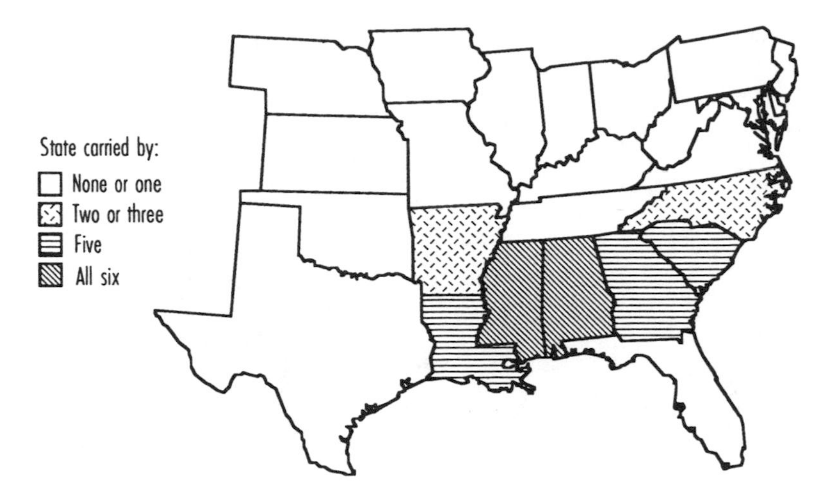

Figure 6. Unusual Voting in Presidential Elections, 1928–1968

Note: Unusual voting is defined as majorities for the following candidates: Al Smith (Democrat, 1928); Strom Thurmond (States' Rights, 1948); Adlai Stevenson (Democrat, 1952, 1956); Barry Goldwater (Republican, 1964); George Wallace (American Independent, 1968).

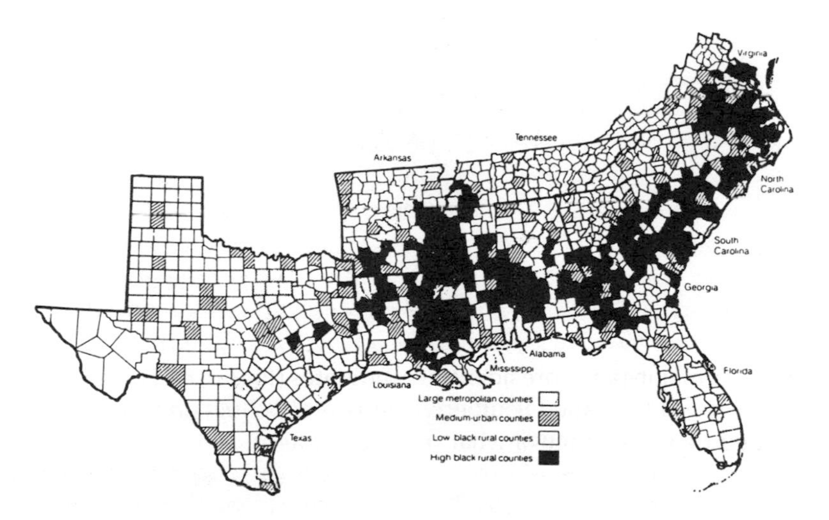

Figure 7. The Demography of the South, 1980

Source: Earl Black and Merle Black, *Politics and Society in the South* (Cambridge: Harvard University Press, 1987), 38.

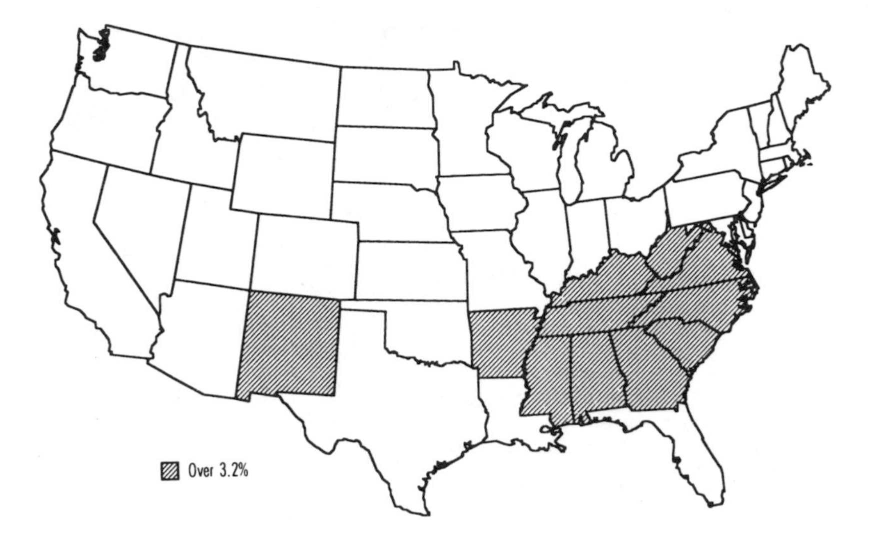

Figure 8. Housing Units without Complete Plumbing, 1980
Source: Data from *Statistical Abstract of the United States, 1982–83,* 755.
Note: Complete plumbing is defined as a flush toilet, a bathtub or shower, and hot and cold running water.

highlands, means that most Southern states are still at the bottom of the U.S. per capita income distribution. (Virginia, Texas, and Florida—barely involved in plantation agriculture, and with little or no mountain population—are exceptions.) This means, in turn, that almost any problem of poor people, or of poor states, can still be used to map the South. Everything from outdoor toilets (Figure 8) to illiteracy (Figure 9) to bad teeth (Figure 10) costs money to put right, and many Southern people and most Southern states still don't have much.

Poverty is bad news, in general, and I certainly don't suggest that we get nostalgic about it, but it has had one or two good points. Burglary rates, for example, are strikingly correlated with states' average incomes—presumably not because rich people steal but because they have more *to* steal—and Figure 11 shows that burglary has been less common in all but the richest Southern states. (A policeman once offered me another explanation. Going in other

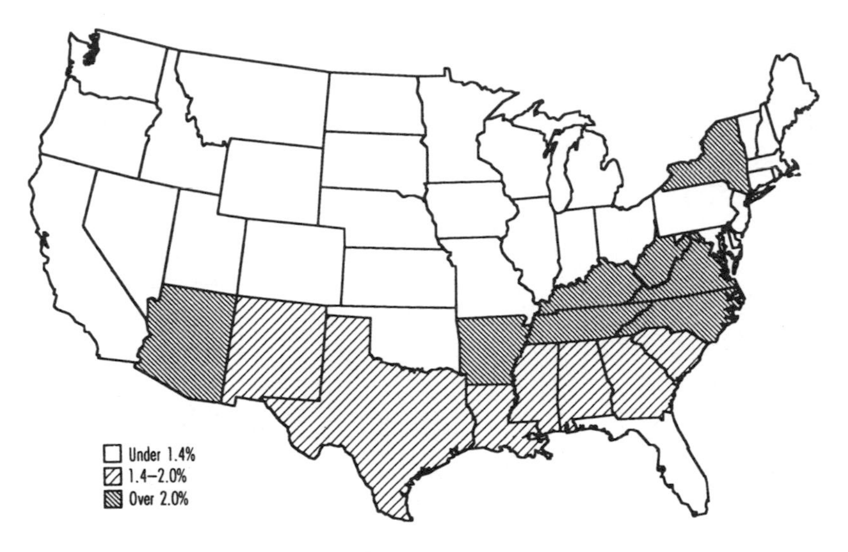

Figure 9. Illiteracy Rates, 1970
Source: Data from *Statistical Abstract of the United States, 1981,* 143.

people's windows is a more dangerous occupation in the South, he argued. "You're more likely to meet something lead coming out.")

In any case, now, the shadow of the plantation is giving way to the light of the "Sunbelt." The South may still be on the bottom of the socioeconomic heap, but the difference between top and bottom is smaller than it used to be. (In a few respects, South and non-South have traded places: the Southern birthrate, for instance, historically higher, is now lower than the national average.) Consequently, those who view the South primarily in economic terms are likely to believe that the region is disappearing. "Southern characteristics" that simply revealed that the South was a poor, rural region are more and more confined to pockets of poverty within the region—or, more accurately, the statistics increasingly reflect the presence of air-conditioned pockets of affluence, particularly in Texas, Florida, and a few metropolitan areas elsewhere. If we map the South with the same criteria people used even fifty years ago, what we get these days looks more like a swiss cheese than a coherent region.

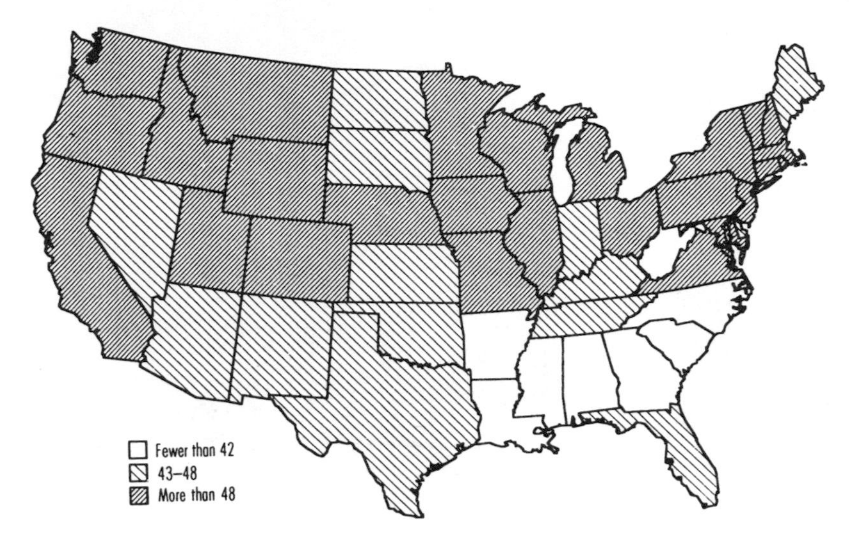

Figure 10. Active Dentists per 1,000 Residents, 1982
Source: Data from *Statistical Abstract of the United States, 1986,* 104.

Figure 11. High Burglary Rates, 1984
Source: Data from *Statistical Abstract of the United States, 1986,* 167.

The Cultural South

But suppose we don't define the South in economic and demographic terms. What if we somehow identify Southerners, and then define the South as where they come from? We could say, for example, that people who eat grits, listen to country music, follow stock-car racing, support corporal punishment in the schools, hunt possum, go to Baptist churches, and prefer bourbon to scotch (if they drink at all) are likely to be Southerners. It isn't necessary that all or even most Southerners do these things, or that other people not do them; if Southerners just do them more often than other Americans, we can use them to locate the South.

Look at the geographical distribution of Baptists, for example (Figure 12). Early on, members of that faith established their dominance in the Southern backcountry, in numbers approached only by those of Methodists. As Southerners moved on to the west and south, they took their religion with them. The map shows a good many Baptists in New York, to be sure, but New York has lots of everything. Not many people live in West Texas, but they're likely to be Baptists. In this respect, the mountain South, too, is virtually indistinguishable from the rest of the region.

And when it comes to Southern music, the mountains and the Southwest are right at the heart of things. Figure 13 shows where country music-makers come from: a fertile crescent extending from southwest Virginia through Kentucky and Tennessee to Arkansas, Oklahoma, and Texas. Musically, what is sometimes called the "peripheral" South is in fact the region's core. The *Deep* South is peripheral to the country-music scene, although it's not a vacuum like New England, and a similar map for traditional black musicians would almost certainly fill some of the gap. Country musicians' origins are reflected in the songs they produce, too: in Figure 14, the size of the states is proportional to the number of times they're mentioned in country-music lyrics. Notice Florida's role as a sort of appendix to the South.

Those lyrics also suggest a regional propensity for several sorts of violence, and FBI statistics show that this isn't just talk. The South has had a higher homicide rate than the rest of the United States for as long as reliable records have been kept, and the mountains and Southwest share fully in this pattern. Southern violence, however,

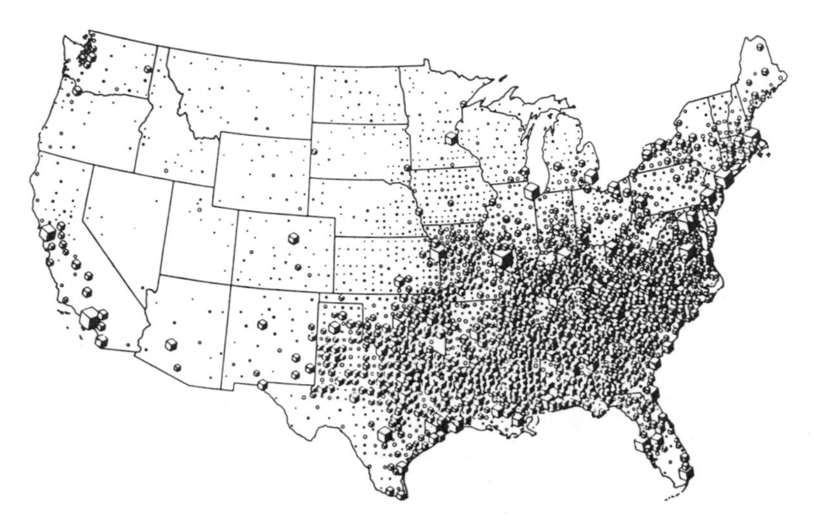

Figure 12. Members of Baptist Churches, 1952

Source: Wilbur Zelinsky, "An Approach to the Religious Geography of the United States: Patterns of Church Membership in 1952," *Annals of the Association of American Geographers* 51 (June 1961): 172.

Figure 13. Birthplaces of Country Music Notables, 1870–1960

Source: George O. Carey, "T for Texas, T for Tennessee: The Origins of American Country Music Notables," *Journal of Geography* 78 (November 1979): 221.

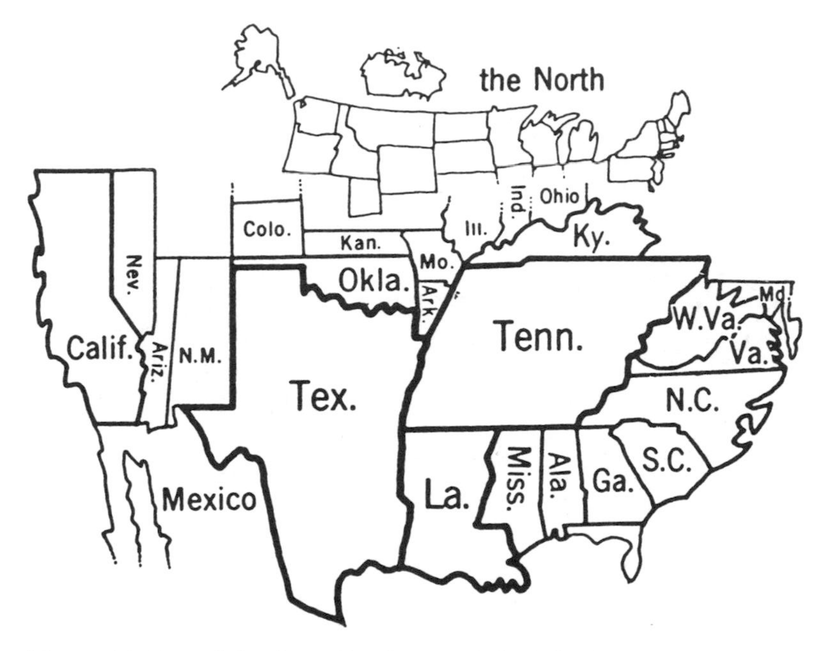

Figure 14. States Mentioned in Country-Music Lyrics
Source: Ben Marsh, "A Rose-Colored Map," *Harper's,* July 1977, 80. Used by permission.
Note: The size of each state is proportional to the number of times it is mentioned.

isn't directed inward. Around the world, societies with high homicide rates tend to have low suicide rates, and the same is true for American states. It very much looks as if there is some sort of trade-off at work. Figure 15 shows where homicide is about as common as suicide—one of the few things the South has in common with New York.

Regional cultural differences are also reflected in family and sex-role attitudes. These differences have even surfaced in the legal system: Southern states were slow to enact women's suffrage; most never did ratify the Equal Rights Amendment; until recently few had state laws against sex discrimination (Figure 16). Southern women have actually been more likely than other American women to work outside the home (they've needed the money more), but most often they've worked in "women's jobs"—as textile operatives or domestic ser-

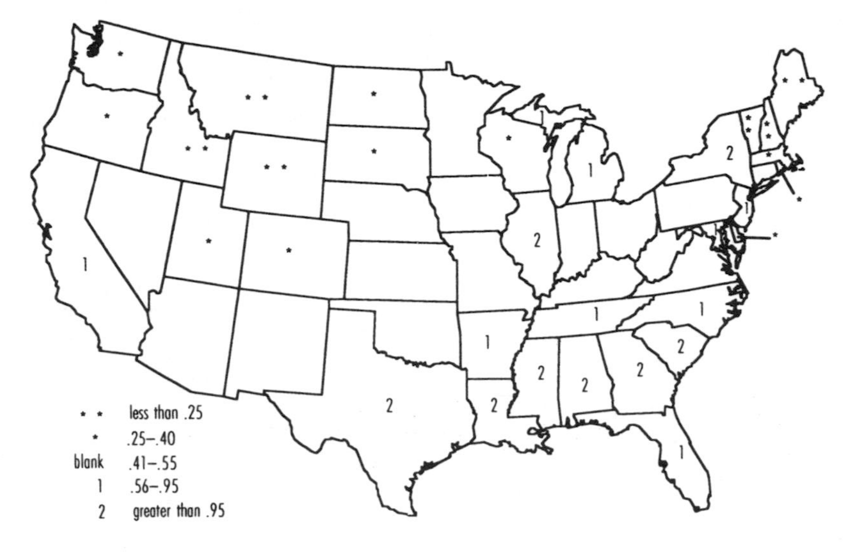

Figure 15. Ratio of Homicides to Suicides, 1983
Source: Data from *Statistical Abstract of the United States, 1987, 77.*

vants, for example. The percentage of women in predominantly male occupations remains lower in the South than elsewhere (Figure 17).

Notice that these characteristics aren't related in any obvious way to the plantation complex. Aspects of culture like diet, religion, sports, music, and family patterns don't simply reflect how people make their livings, or how good a living they make. To a great extent, they're passed on from generation to generation within families. Usually, when families move they carry these patterns with them.

That's why these values and tastes and habits are found in the Appalachians and the Ozarks, and in most of Texas and Oklahoma. Those areas were marginal, at best, to the plantation South, but they were settled by Southerners, and by measures like these they are quite comfortably Southern. Mapping things like this makes it easy to figure out who settled most of Missouri, too, as well as the southern parts of Illinois, Indiana, and Ohio. And many of the same features can be found in scattered enclaves of Southern migrants all around the United States—among auto workers in Ypsilanti, for instance, or the children and grandchildren of Okies in Bakersfield.

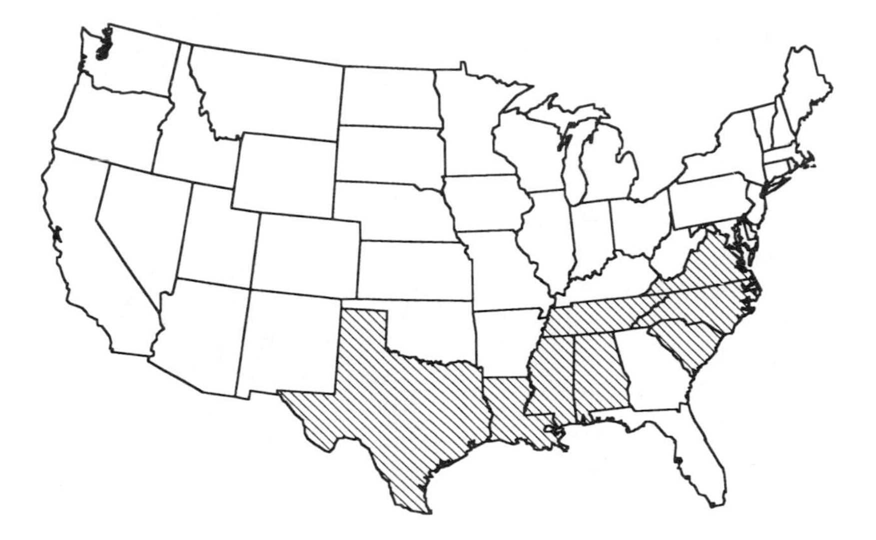

Figure 16. No State Law against Sex Discrimination, 1972

And we can't expect the demise of the plantation to make these characteristics go away. So if we define the South as a patch of territory somehow different from the rest of the United States because it is inhabited by people who are different from other Americans, we still have a great deal to work with.

Indeed, we have new things to work with all the time. We need to recall that country music came of age only with the phonograph, and NASCAR only with the high-performance stock car. Consider also Figure 18, which locates colleges and universities that publish their own sports magazines. Southern institutions of higher learning have seldom been on the cutting edge of innovation, but they seem to be out front on this one.

Southern Identification

I suggested earlier that we can look at the South, not as just a distinctive economic or cultural area, but as the home of people somehow bound together by ties of loyalty and identification. Clearly,

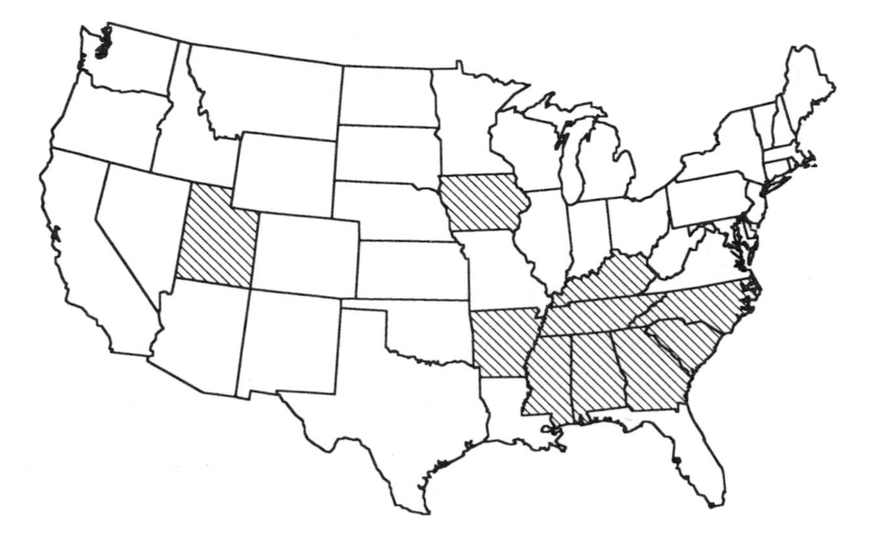

Figure 17. Low Percentage of White Women Employed in Traditionally White Male Occupations, 1985

Source: Data from Southern Labor Institute.
Note: The bottom ten states are shaded.

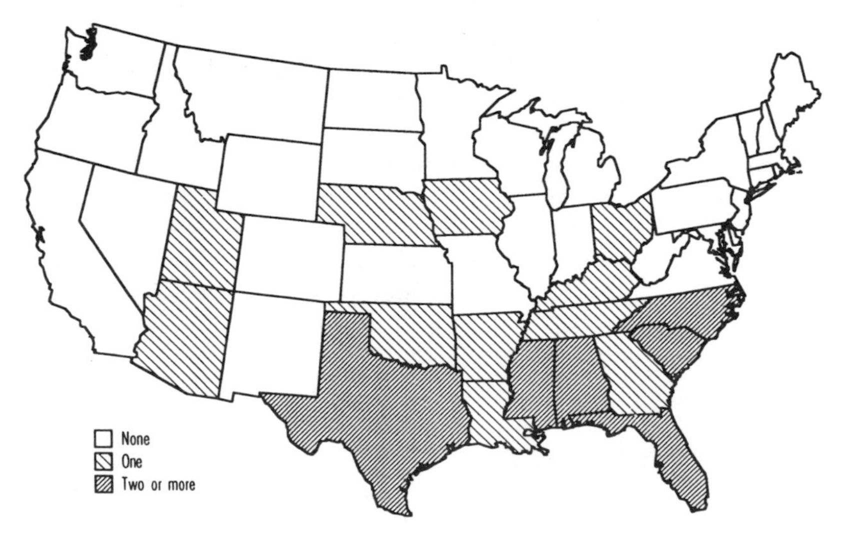

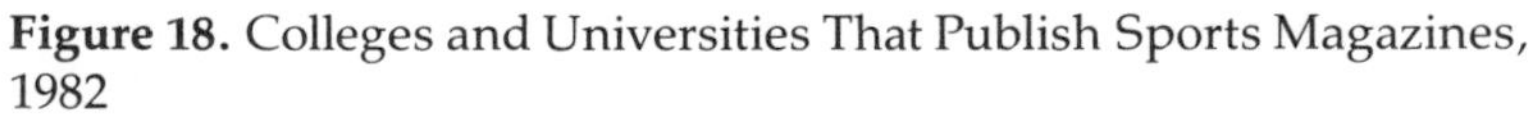

Figure 18. Colleges and Universities That Publish Sports Magazines, 1982

Source: Data from *Chronicle of Higher Education,* September 15, 1982, 17.

the South has been a "province," in Josiah Royce's sense of that word: "part of a national domain which is, geographically and socially, sufficiently unified to have a true consciousness of its own unity, to feel a pride in its own ideals and customs, and to possess a sense of its distinction from other parts of the country."

Not long ago, the regional patriotism of most white Southerners was based on the shared experience of Confederate independence and defeat. There are still reminders of this past in the South's culture and social life. Figure 19, for example, shows where to find chapters of the Kappa Alpha Order, a college fraternity with an explicitly Confederate heritage.

For many, the word *Dixie* evokes that same heritage, and Figure 20 shows where people are likely to include that word in the names of their business enterprises. Notice that the Appalachian South, which wasn't wild about Dixie in 1861, still isn't. Now the Southwest, too, has largely abandoned Dixie (turn about: the Confederacy largely abandoned the Southwest, once). Most of Florida would probably be gone as well if there were no Dixie Highway to keep the word in use. Even in the city of Atlanta, Dixie seems to be gone with the wind—or at least on the way out. Only in what is left of the old plantation South is Dixie really alive and well.

As a basis for identification, obviously, symbols of the Confederate experience necessarily exclude nearly all black Southerners, as well as many Appalachian whites and migrants to the region who are so recent that they haven't forgotten that they're migrants. Fortunately, however, regional loyalty can be based on other things, among them cultural differences like those we've already examined.

We can ask, in other words, not "where do people display Southern ways?" but "where do people assert the superiority of Southern ways?" Figure 21, for example, shows where people are likely to say that they like to hear Southern accents, prefer Southern food, and believe that Southern women are better looking than other women. (The Gallup Poll hasn't asked these questions lately, so the data are a little old, but I doubt that the patterns would be much different now.) The South defined this way naturally coincides pretty well with the area where one is actually apt to encounter Southern accents, Southern food, and Southern women—a bigger region than what remains of the Confederate South, just as the cultural South extends well beyond the domain of the old plantation system.

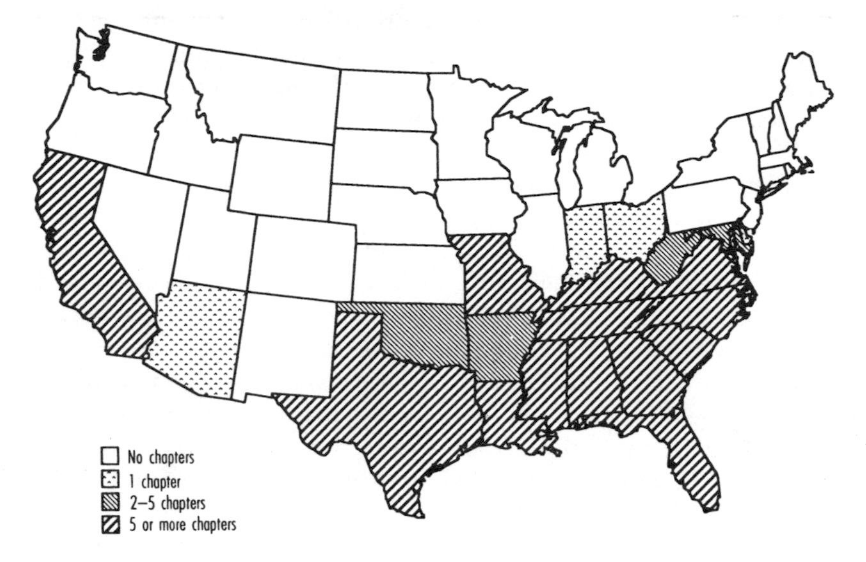

Figure 19. Chapters of the Kappa Alpha Order, 1988
Source: Data from Upsilon of Kappa Alpha.

Regional Institutions

Regional institutions play a part in sustaining the South, as both idea and reality, tying the region together economically and socially and contributing to a sense of distinctiveness and solidarity. Here, too, we find a close analogy to the life of American ethnic groups. Like some of those groups, Southerners have their own social and professional organizations, organs of communication, colleges and universities, and so forth. In fact, they probably have more of them now than they ever did before. When Karl Marx said scornfully of the Confederacy that it wasn't a nation at all, just a battle cry, he was referring to the absence of this sort of institutional apparatus, and until recently the South couldn't *afford* much in the way of regional institutions.

But now the Southern Historical Association, the Southern Railway, the Southern Baptist Convention, the Southern Growth Policies Board—these and other, similar institutions establish channels

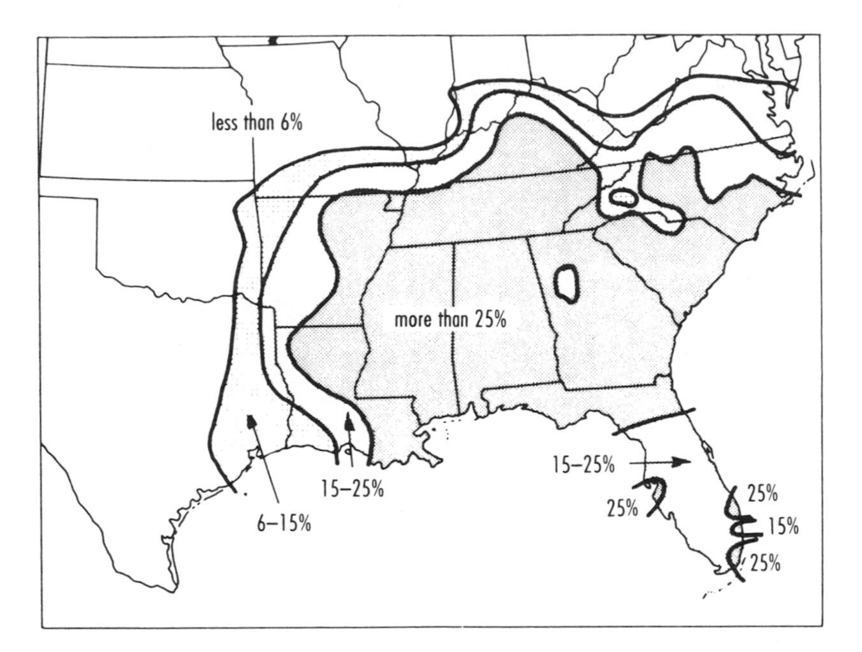

Figure 20. "Dixie" Listings as Percentage of "American" Listings in Telephone Directories, ca. 1975

Source: J. S. Reed, "The Heart of Dixie: An Essay in Folk Geography," *Social Forces* 54 (June 1976): 932.

of communication and influence within the region, making it more of a social reality than it would be otherwise. At the same time, even the names of organizations like these serve to reinforce the idea that the South exists, that it means something, that it is somehow a fact of nature.

Southern Living magazine, for instance, implies month after month that there is such a thing as Southern living, that it is different and (by plain implication) better. Figure 22 shows where that message falls on fertile ground. Notice that Floridians are relatively uninterested in it. So are Texans, despite heroic efforts by the magazine (including a special Southwestern edition). Here we see plainly a development that regional sociologists were predicting fifty years ago, something maps of regional culture and regional identification

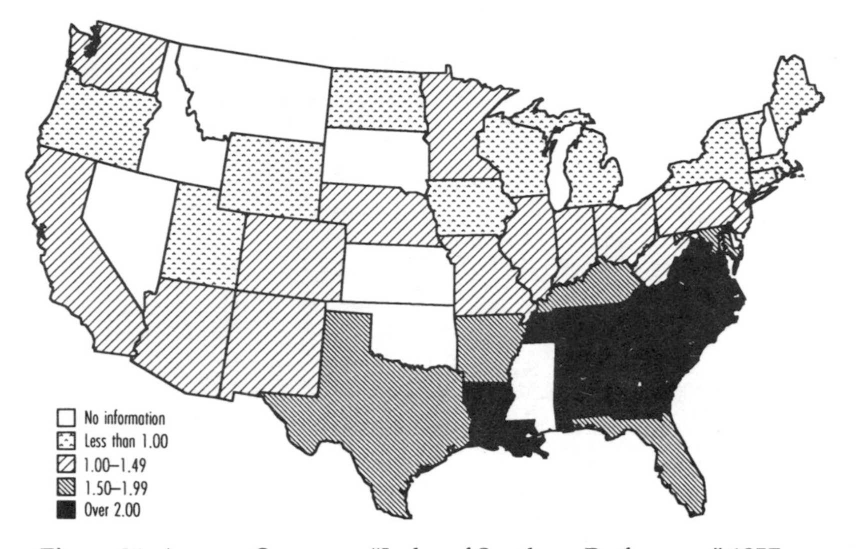

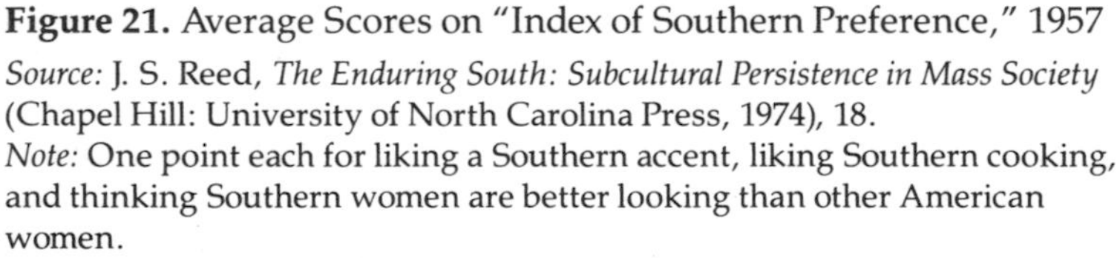

Figure 21. Average Scores on "Index of Southern Preference," 1957
Source: J. S. Reed, *The Enduring South: Subcultural Persistence in Mass Society* (Chapel Hill: University of North Carolina Press, 1974), 18.
Note: One point each for liking a Southern accent, liking Southern cooking, and thinking Southern women are better looking than other American women.

only hint at: the bifurcation of the South into a "Southeast" centered on Atlanta and a "Southwest" that is, essentially, greater Texas. (Texas has its own magazines.)

We find something similar when we look at one of the South's regional universities. The University of North Carolina, in Chapel Hill, has long been a center for the study and nurture of Southern culture. It has also helped to educate a regional elite. Figure 23 shows where an appreciable percentage of all college graduates are Chapel Hill alumni. Tar Heels are thick on the ground throughout the southeastern states, but (aside from some brain drain to the New York City suburbs) that's the only place they're thick on the ground. In particular, Chapel Hill has little market penetration west of the Mississippi. (Texas has its own universities.)

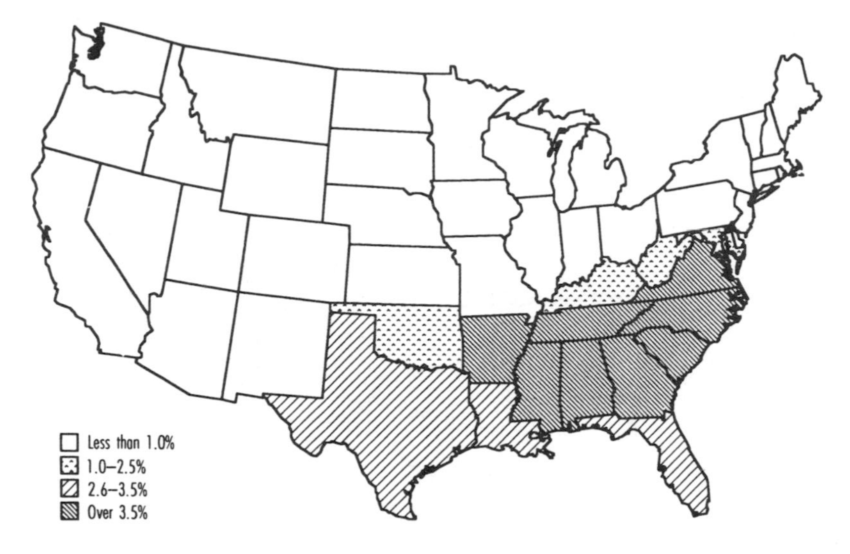

Figure 22. *Southern Living* Readers as Percentage of White Population, 1981

Source: Data from Marketing Department, *Southern Living.*

So Where Is It?

So where is the South? Well, that depends on which South you're talking about. Some places are Southern by anybody's reckoning, to be sure, but at the edges it's hard to say where the South is because people have different ideas about *what* it is. And most of those ideas are correct, or at least useful, for some purpose or other.

The South is no longer the locus of a distinctive economic system, exporting raw materials and surplus population to the rest of the United States while generating a variety of social and economic problems for itself. That system's gone, and good riddance. Some of its effects still linger, though, and a few—such as a genuinely biracial population—will be with us for the foreseeable future.

The South is also set apart by its people and their distinctive ways of doing things. Mass society has made some inroads, but Southerners still do many things differently. Some are even inventing

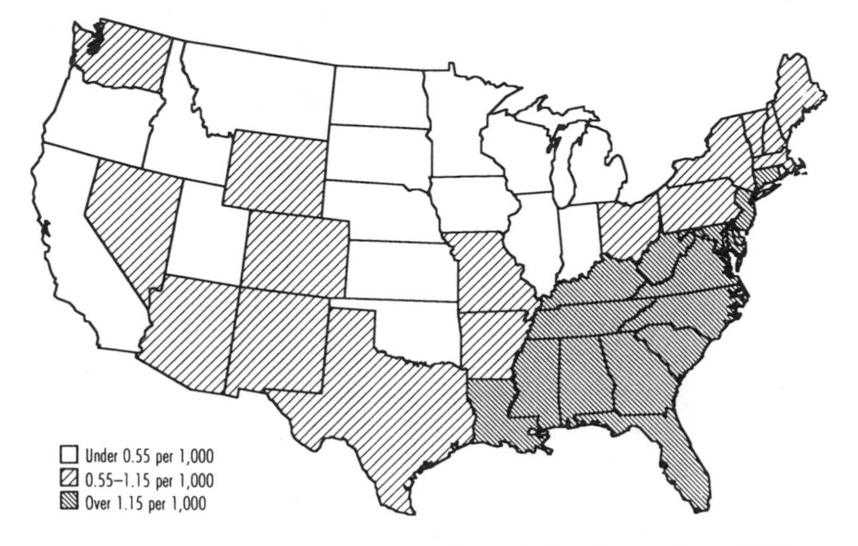

Figure 23. Alumni of the University of North Carolina at Chapel Hill as Estimated Proportion of All Residents with 1–4 Years of College, 1985
Source: Data from Alumni Office, University of North Carolina at Chapel Hill.

new ways to do things differently. And the persistence of the cultural South doesn't require that Southerners stay poor and rural. Indeed, poor folks can't afford some of its trappings: bass boats and four-wheel-drive vehicles, for instance.

Because its history and its culture are somewhat different from the run of the American mill, the South also exists as an idea—an idea, moreover, that people can have feelings about. Many are fond of the South (some even love it); others have been known to view it with disdain. In either case, the South exists in people's heads and in their conversations. From this point of view, the South will exist for as long as people think and talk about it, and as for its boundaries—well, the South begins wherever people agree that it does.

Finally, the South is a social system, perhaps more now than ever before. A network of institutions exists to serve it, and an ever-increasing number of people have a crass, pecuniary interest in making sure that it continues to exist. Here, the brute facts of distance and diversity conspire to reduce the South to a southeastern core.

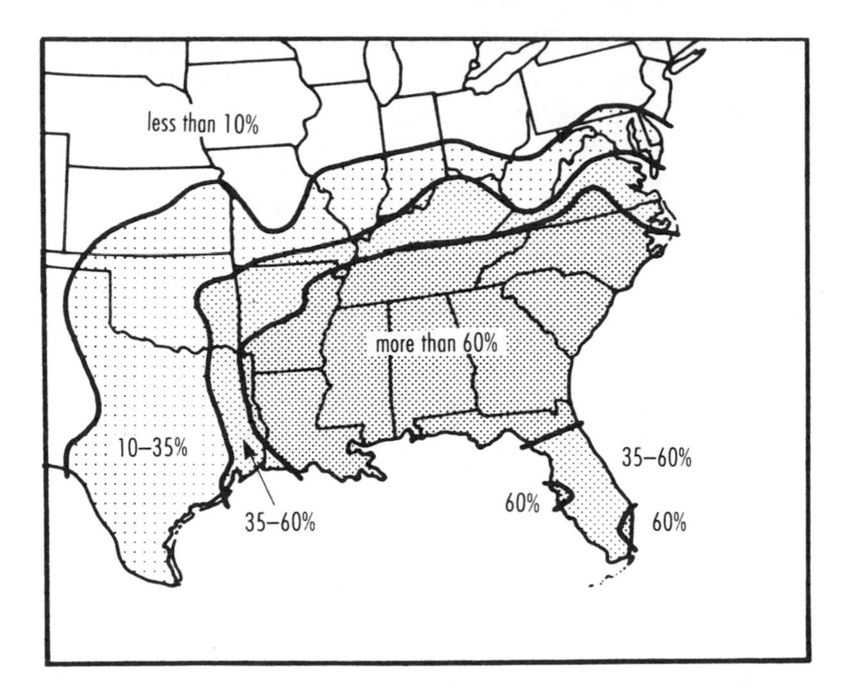

Figure 24. "Southern" Listings as Percentage of "American" Listings in Telephone Directories, ca. 1975

Source: J. S. Reed, "The Heart of Dixie: An Essay in Folk Geography," *Social Forces* 54 (June 1976): 929.

Given all these different Souths, obviously, we can't just draw a line on a map and call it the South's border. As Southerners are fond of saying: it depends. But, what the hell, if I had to do it, my candidate would be the line in Figure 24 that shows where "Southern" entries begin to be found in serious numbers in urban telephone directories (the one at 35 percent).

The South below that line makes a lot of sense. It includes the eleven former Confederate States, minus all of Texas but the eastern edge. It also includes Kentucky, which had a wishful star in the Rebel flag, but not Missouri, which did too. A corner of Oklahoma makes it in as well: we get Muskogee.

Figure 24 shows variation within the South that also makes sense.

It follows some of the stress lines we've already looked at. Kentucky and much of Virginia, East Texas and part of Arkansas, most of peninsular Florida—all these areas on the edges of the South are less "Southern" than the regional heartland, by this measure as by others we've examined. On the other hand, a Southern sphere of influence takes in Maryland, West Virginia, Oklahoma, much of Texas, and the southern parts of the states from Ohio west to Missouri. Few would include most of these areas in the South proper, but fewer would deny their Southern cultural flavor.

This one statistic indicates the presence of the sort of regional institutions I mentioned earlier, as well as the kind of regional enthusiasm that leads an entrepreneur to call a newsstand, say, the Southern Fruit and News. It shows, that is, where the idea of the South is vital, where its social reality extends, or both.

In other words, if you want to know whether you're in the South, you could do worse than to look in the phone book.

Southerners as an American Ethnic Group

■ If the ancestors of most present-day Southerners had had their way, there would be no question of whether they should be treated as an *American* ethnic group. Southern whites remained—as Southern blacks became—Americans against their will, a fact that has a great deal to do with their persistence as an identifiable group.

Southerners, like members of immigrant ethnic communities, belong to a group defined by a historical experience, in which membership is ordinarily inherited but rests ultimately on a combination of individual identification and acceptance or categorization by others. Like other ethnic groups, Southerners have differed from the national norm: they have been poorer, less educated, more rural, occupationally specialized. They also differ culturally in important respects, and their political behavior has been distinctive. Although Southerners are not often identifiable by name or appearance, their accent usually serves as an ethnic marker. They have been stereotyped by other Americans and, indeed, are usually willing themselves to generalize about their differences from their countrymen. The extent of discrimination against them—in the appointment of Supreme Court justices, for example—is a matter of opinion, but it is significant that it is even that.

The analogy between Southerners and immigrant ethnic groups is complicated by the fact that, for Southerners, the "old country" is part of the United States, and most still live there. Moreover, the South has always received migrants from the rest of the country and turned many of them, and more of their children, into Southerners themselves. The South has been defined in various ways, but it almost always includes the eleven former Confederate states. Most definitions would exclude parts of these states—west Texas and northern Virginia, for example, and southern Florida, if there were anywhere else to put it—and include at least parts of Kentucky and sometimes Oklahoma and the various border states. However the region is defined, though, it is clear that the term *Southerners* does not mean simply residents of the South, but refers rather to people who somehow identity themselves with the region, whether they live there or not.

Identification

Southern identification—a sense of the South as an entity over and above the states and localities that make it up, and some sense of patriotism toward that entity—was shaped by the sectional conflict of the early nineteenth century, primarily among whites from those areas where slavery and the plantation system were well established. As agriculturists, they often found their interests to be different from those of the Northeast; as residents of a biracial region with a slave system, their interests were not those of the agricultural Midwest. From the foundation of the Republic, the representatives of the Southern states found themselves allied in defense of agriculture and of slavery, and regional conflict over economic and racial issues did not end with the Civil War. The war itself, with its legacy of defeat, occupation, and subordination, gave the former Confederates another common basis for identification, a distinctiveness based on history rather than on current circumstances. They had not only their own flag, anthem, and holidays, but a heritage of economic—and, allegedly, cultural and moral—inferiority to set them apart from other Americans.

Even today, identification with the South appears to have something to do with ancestral, if not one's personal, sympathy with the Confederate cause. A 1971 study in North Carolina found both blacks and Appalachian whites (from areas largely pro-Union in the Civil War) less likely than other North Carolinians to regard themselves as Southerners, some explicitly rejecting the label on nationalist (that is, Unionist) grounds. Similarly, another study showed that regional identification among Southern whites was strongest among those from the conventionally defined Deep South, weaker among those from Appalachia and more recently settled areas of Florida and the Southwest, and weakest of all among migrants to the South. The same study found very little relation between thinking of oneself as a Southerner and support for racial segregation. The importance of ancestral commitment is underlined by the results of another study, which showed anti-Southern sentiment among non-Southern whites to be most common in New England, particularly among Republicans there.

Nevertheless, majorities of Southern blacks and Appalachian whites and a substantial minority of migrants to the South do claim to be Southerners. Whether—or, rather, when and by whom—this claim is accepted as valid is another question, but it is clear that there is more to the matter of Southern identification than having fought the Yankees and lost, several times. The importance of that experience seems to be diminishing: many residents of the South whose connection with the Lost Cause is complicated or nonexistent have begun to assert that they, too, are Southerners. Whether this assertion springs from a shared history of a more complex sort, from a shared cultural style, from current and future common interests, or from some combination of these, it is obvious that *Southerner* does not mean the same as *ex-Confederate.*

Whatever its basis, the level of identification among Southerners is surprisingly high. Data collected in the 1960s from native-born white residents of the eleven former Confederate states show that, on a commonly used index, the group identification of Southerners surpassed that of Roman Catholics with others of their religion and that of union members with other unionists, and approached the levels of identification displayed by Jews and blacks.

Distinctiveness

Although Southerners retain a strong sense of themselves as different from other Americans, in some important respects they differ less than before. Census data deal with residents of the South rather than with self-identified Southerners but make it clear that with each passing decade Southerners are more and more "American" in the ways they earn their living, in the return they receive from that undertaking, in the educational background they bring to it, and in the settings in which they do it.

The South was always an agricultural region, and, at least after 1865, it was a poor region as well. As late as the 1930s two-thirds of its people were rural, and half were employed directly in agriculture; most of the remaining "urban" third lived in small towns and villages. Per capita income was roughly at the level used today to distinguish developed from underdeveloped countries, and the usual implications for social welfare, public health, and education followed. Since then, however, the South, like the rest of the United States, has become an urban, industrial, and comfortable (if not affluent) society. By 1980 two-thirds of the population was urban, and several Southern cities were among the nation's largest and fastest growing. Less than a tenth of the population was employed in agriculture, and Southern incomes, though still below the national average, were increasing steadily and closing the gap. Southerners by the 1970s were as likely as other Americans to have had some college education; the region's birthrate had dropped to below the national average; a pattern of net migration from the region, established for over a century, had been reversed—more blacks and whites were moving to the South than from it. The economic and social problems that affected the region as a whole in the 1930s remained for the most part only in pockets, notably in parts of Appalachia and the old Cotton Belt. Economically and demographically the region was becoming much like the rest of the United States, and Southerners were becoming much like other Americans.

But there are other ways in which Southerners have been seen as different from other Americans. They are seen, and see themselves, as less energetic, less materialistic, more traditional and conventional, more religious and patriotic, more mannerly and hospitable. Many of these cultural generalizations existed in much their present form well before the Civil War, and it is interesting that Americans' stereotypes of Southerners resemble in some ways the traditional white stereotype of blacks. In general, non-Southerners' images of Southerners are about the same as Southerners' stereotypes of themselves, although the evaluation may differ. (*Easygoing* and *lazy*, for example, express much the same perception.) Both see Southern culture as, in many ways, simply the reverse of "American" culture.

A study of Southern college students, for instance, found that their image of "Northerners" was almost identical with their general image of "Americans," but their view of their own group had very little in common with either.

Perhaps not surprisingly, though, Northerners' images of Southerners seem to be less differentiated than Southerners' perceptions of their own group. A study in the urban Midwest by Lewis Killian, for instance, reported that the natives stereotyped migrants from both the Appalachian and the lowland South indiscriminately as "hillbillies," ignoring a distinction that has been fairly important in the South. Similarly, Peter Gould and Rodney White reported, in a study of the "mental maps" of college students, that non-Southerners tended to regard the South as a (rather unappetizing) whole, ignoring differences that Southern students perceived between, for instance, North and South Carolina, or Alabama and Mississippi.

The accuracy of the folk generalizations embodied in the common regional stereotypes remains undetermined, but large cultural differences between Southerners and other Americans seem to persist, and in this respect the stereotypes appear to bear some relation to the facts. Research on this subject generally has been confined to whites, but there is little reason to suppose that regional differences are not much the same among blacks, allowing for the fact that many non-Southern blacks are either Southern-born or have Southern parents.

A series of studies in the 1960s, by Norval Glenn and his colleagues, examined a variety of cultural questions taken from a number of national public opinion polls and compared the responses of white Americans from each of four major regions: the Southern, Northeastern, Central, and Western states. The South was by far the most distinctive of these "regions," and the differences between white Southerners and other white Americans were substantial—larger, on the average, than those between rural and urban people, Protestants and Roman Catholics, or manual and nonmanual workers, and about the same as those between blacks and whites. Some of the differences between Southern respondents and others had decreased in the previous generation, but the average of all the differences had not. If anything the average had increased, and it appeared to be larger among young people than among older ones.

My own study, *The Enduring South,* found substantial differences between white Southerners and other white Americans in religious beliefs and practices (even comparing only Protestants), in attitudes toward the family and the local community, and in attitudes and reported behavior involving violence and the private use of force. (This last difference is also reflected in the historically higher rate of homicide and some other—but not all—violent crimes in the South.) In general, these differences had not decreased during the generation or so covered by the study, and they could not be explained in any obvious way by regional economic and demographic differences.

Although the racial attitudes of white Southerners are beginning to resemble those of other white Americans, in other respects their political attitudes have become even more distinctive over the past generation. Politically, Southerners have become more conservative, particularly with regard to issues concerning foreign policy and the size of the federal government, and the gap between the attitudes of Southerners and those of other Americans has widened.

There is no reason to suppose that the South's cultural distinctiveness is going the way of its economic and demographic peculiarity. The increasingly common experiences of urban life, education, travel outside the South, and exposure to the national mass media weaken some components of traditional Southern culture (ethnocentrism and religious fundamentalism, for instance), and urban, educated Southerners in nonmanual occupations are increasingly "assimilated" in these respects. (According to Peter Trudgill, these same middle-class elements seem also to be acquiring the postvocalic *r*, the absence of which has heretofore marked most Southern accents.) But in other ways in which Southerners differ from the national mainstream—in political attitudes, some aspects of religious behavior, and attitudes toward violence, for instance—regional differences are at least as large among urban people as among rural ones, among the educated as among the uneducated, and among business and professional people as among blue-collar workers.

Political Behavior

Nowhere has Southerners' distinctiveness been more apparent than in their political behavior—most strikingly, of course, in their attempt to secede (although that attempt was opposed by many Appalachian whites, and Southern blacks did not have a voice in the matter). Although 11 percent of white North Carolinians polled in the 1970s said that they would still favor an amicable secession, most Southern whites since 1865 have been content to pursue what they viewed as their region's interests within the national political system.

From the end of Reconstruction until the 1960s, the Democratic "solid South" was almost a given in American political life. By disfranchising blacks (and many poor whites as well), the political leadership of the region assured the Democratic party of dependable majorities and became what seemed an indispensable part of national Democratic coalitions—first, the nineteenth-century one of "rum, Romanism, and rebellion"; later, Franklin D. Roosevelt's New Deal alignment. The region's strength in the party usually gave its representatives at least veto power over Democratic presidential nominations, and the combination of one-party politics and the seniority system in Congress gave Southern senators and congressmen power out of proportion to their numbers.

Although this comfortable alliance showed signs of strain as early as 1928, the first major break occurred twenty years later, when many Southern whites believed Harry Truman to be insufficiently committed to racial segregation. Especially in the Deep South, voters defected to support Strom Thurmond, candidate of the States' Rights (Dixiecrat) party. In 1952 and 1956 these same voters returned to their habitual Democratic pattern (and in fact gave Adlai E. Stevenson his only majorities), but in 1964 the pattern was almost wholly reversed: the Deep South voted solidly for a Republican, Barry Goldwater, who had indicated his opposition to civil rights legislation. (As in 1952 and 1956, the loser in 1964 carried only states in the South.) In 1968 several of these same states again turned to a regional third-party candidate, George Wallace, who alleged that there was not "a dime's worth of difference" between the two major parties.

Although the states that delivered majorities for these third-party and unpopular major-party candidates were primarily in the Deep South, similar tendencies can be seen among white voters in other Southern states, especially those in areas with large black populations. But in the states of the peripheral South, for various reasons, issues other than race—Al Smith's religion, Dwight Eisenhower's personal appeal, or simple economic interest—sometimes played a part in determining elections.

The Voting Rights Act of 1965, by effectively guaranteeing the right of Southern blacks to vote, has changed the face of Southern politics, but it would be premature to say that those politics have been "nationalized." Although the outcome of Southern elections may conform increasingly to the national pattern, the ingredients remain somewhat different. In 1976, for example, all the Southern states delivered majorities for Jimmy Carter, the Democrat, but a majority of the region's white voters once again supported the loser: Carter won most Southern states because of the votes of blacks. And had Carter not been a Southerner himself, his vote among Southern whites presumably would have been even smaller.

Religion

If the political solid South is breaking up, the religious solid South—which antedates it—persists. About 90 percent of the region's whites identify themselves as Protestants; about half of those are Baptists, and a sizable minority are Methodists. An even higher proportion of Southern blacks are also Baptists or Methodists, the great majority of them in all-black denominations. Even these figures understate the religious homogeneity of most Southern communities, since non-Protestants are geographically concentrated within the region.

The fact that most Southerners belong to low-church, Evangelical Protestant denominations has affected the region's life in ways both obvious and subtle. The democratic polities and anti-authoritarian traditions of these churches, for example, may help to explain why black churches and church people were conspicuous in the struggle for black civil rights, while some white churches and church people were equally active in the opposition.

Southerners are orthodox in their religious beliefs and vigorous in support of their churches. Eighty-six percent of white Southern Protestants told the Gallup Poll, for example, that they believed in the Devil, compared to 52 percent of white Protestants elsewhere. Other studies of belief show much the same pattern of regional difference. Southern white Protestants are more likely than other white

Protestants to attend church, to listen to religious programs on the radio, to name a religious figure (usually Billy Graham) as the man they most admire, and to believe that regular church attendance is a necessary part of the Christian life. (They are even more likely to believe the last than Roman Catholics.)

Church attendance within the South also differs from patterns in the rest of the country. Outside the South, urban blue-collar Protestants are among the population groups least likely to attend church, much less likely to do so than relatively uneducated rural people. But in the South the urban working class is as likely to attend church as are similarly educated rural folk. Rural-to-urban migration in the South apparently has not disrupted established patterns of religious practice. Moreover, educated, urban business and professional people are among the most churchgoing of Southerners; they are more likely to attend church services (or at least to say they do) than their white-collar employees—a reversal of the non-Southern pattern. The leadership of the New South, like that of the Old, apparently regards regular churchgoing as part of its role.

A few of the more distinctive correlates of Southern religiosity—such as opposition to the sale of alcohol and support for Sunday blue laws—are waning, but much remains. For some time to come, Southerners will be characterized not only by religiosity but by religiosity of a distinctive kind.

Literature and Music

The verbal emphasis of the Southern religious tradition may help to explain the remarkable Southern contribution to American literature. For reasons not clearly understood, many of America's outstanding writers in this century have been Southerners—William Faulkner, Thomas Wolfe, and Flannery O'Connor, to mention only three. A regional fascination with words may also be evident in the prominence of Southern historians, literary critics, jurists, and journalists. This record has given the lie to H. L. Mencken's picture of the South as a cultural wasteland, but the absence of any remotely comparable excellence in, say, the visual arts does require an explanation, and that may also lie in the South's religious tradition.

In another respect Southerners have contributed disproportionately to American cultural life: the region has been a great—probably the greatest—seedbed of American folk and popular music. The relations of jazz to New Orleans, of the blues and rock music to Memphis, and of "hillbilly" and country music to Nashville are well known, and lesser centers have also played a part. What has come to be regarded as distinctively American music has largely been Southern music, created by black and white Southerners drawing on folk traditions developed and maintained, in part, in their churches.

Like Coca-Cola, Holiday Inns, and Kentucky Fried Chicken, however, the South's literary and musical contributions to American life have been adopted by the nation as a whole and can no longer be viewed solely or even primarily as features of an ethnic "Southern" culture. Some of them, indeed, never were that: sadly, modern Southern writers have usually found their publishers and most of their readers outside the region. In literature and music as in economy and demography, South and non-South are becoming more alike. But there is some question as to who is assimilating whom.

Ethnic Institutions

As a regional group, Southerners have developed economic, cultural, and political institutions that exist to serve residents of the region rather than ethnic Southerners, but the overlap between the two categories means that these institutions often function in a quasi-ethnic fashion, without imposing tests of loyalty or ancestry on their members or employees, and, indeed, without recognizing that they are doing so. Similarly, at the community level many local institutions operate in a distinctively Southern way—a fact that would be recognized more often if there were more non-Southerners present to call attention to it. Studies have found regional differences, for example, in the values expressed by daily newspapers, Methodist clergymen, and public school teachers. To be sure, these differences reflect the fact that most journalists, clergymen, and teachers in the South are themselves Southerners, but the point remains that the press, pulpit, and schoolroom in the South have by no means been entirely assimilated to "national" models and continue to define, interpret, and reinforce a regional culture.

Besides the institutions that the South has as an American region and that Southern communities have as American communities, Southerners have a number of institutions analogous to those of other American ethnic groups. These include filiopietistic, ancestry-based organizations like the United Daughters of the Confederacy; militant ethnic defense associations, of which the Ku Klux Klan is only the best known; college fraternities, such as Kappa Alpha; and even a university (the University of the South, at Sewanee), which, like some other ethnic universities, never quite fulfilled its founders' expectations. Many of these organizations—the university is an exception—are devoted to a version of Southern culture that includes white supremacy, at least as an implication, and many adopt the imagery and mythology of the Confederate cause, thereby effectively excluding much of the population of the South from association with them. A related development since the desegregation of Southern public schools has been the growth of private "segregation academies," often under the auspices of Protestant churches whose opposition to parochial education was formerly unrelenting. These schools, too, often preserve a version of Southern culture that includes (but is certainly not limited to) white racism.

A more comprehensive vision has been represented by a few self-consciously sectional organizations like the Lamar Society—a progressive, "good government" organization on the model of Common Cause—that take pains to be biracial, and by some straightforwardly commercial enterprises, such as *Southern Living* magazine, that market their version of the good life without regard to the race or creed of their customers. By and large, members of white minority groups in the South—Jews and Roman Catholics, in particular—have chosen to assimilate with the dominant regional group while preserving their separate religious or ethnic identity, and they have been surprisingly successful in this maneuver. It remains to be seen to what extent Southern blacks and the new wave of migrants to the South will elect, and be allowed to pursue, a similar course.

Migrants to and from the South

A particularly interesting group, but one about which little is known, is that composed of migrants to the South. It appears that many are Southern-born, returning to their home region now that jobs and welfare benefits are becoming as attractive there as elsewhere. Most of these migrants have maintained ties to the South and should have few difficulties of assimilation, although their attitudes and expectations have, of course, been affected by their residence outside the South.

Migrants to the South who have never lived there before are a different matter. Their direct effect on the culture of the region is probably reduced by their concentration in the larger cities and in such "non-Southern" enclaves as Oak Ridge, Tennessee, and Huntsville, Alabama, where their contact with the native-born is often minimal. Although such enclaves have become a factor in the politics of some Southern states, they have little effect on everyday life elsewhere in the region, nor does the regional setting greatly affect life within them. Moreover, some evidence suggests that migrants are self-selected to begin with—that is, they are more "Southern," culturally, than non-Southerners who stay at home. Those who find themselves in situations where assimilation is desirable should have little trouble fitting in, as previous generations of migrants have done.

The situation of migrants from the South to other regions is similar: they are generally less "Southern" than those who do not leave the region. (Thus their departure may contribute to the maintenance of cultural distinctiveness in the South.) Among these migrants, too, it is likely that regional identification seldom persists beyond the first generation—or even that long if the regional accent, almost the only ethnic marker, is lost—unless they settle in Southern enclaves outside the South.

Many of course have settled in such communities, in the urban Northeast and Midwest and on the West Coast. Because such enclaves —especially of Southern blacks and working-class whites—tend to be regarded as problems, they have been studied and analyzed. Here, it seems, the analogy between Southerners and immigrant ethnic groups is quite close: in these predominantly working-class areas, churches, clubs, and taverns play a role in sustaining group identity and cultural life, reinforced by frequent visits to relations "down home"; characteristically Southern institutions, loyalties, and behavior patterns abound. Various studies have related the presence of Southern migrants to the transformation of the Southern

Baptist Church into a national (indeed, international) denomination, to votes for Carter in the 1976 presidential election, and to state-to-state variations in the homicide rate. Nostalgia for the South is a frequent theme in country music, and Bakersfield, California, is nicknamed "Nashville West"—testimony to its role in sustaining the culture of second-generation migrant Southerners.

Those who like their boundaries well defined should not attempt to talk about Southerners—a "quasi-ethnic" group that cuts across conventional ethnic distinctions, based on some ancestral or current connection with a region that is itself indeterminate, involving a sense of identification that, for many, depends largely on their immediate social setting. But, when all is said and done, this ill-defined mass of Americans has resisted and continues to resist the assimilating effects of powerful twentieth-century political, economic, and social forces, and maintains, in the face of those forces, a sense of its own distinctiveness.

The South's Mid-Life Crisis

■ "Mid-life crisis." Why that phrase from middle-brow pop psychology? What does that sort of bi-coastal psychobabble have to do with the South?

You might well ask. After all, one thing some of us like about it here is that Southerners spend less time than other Americans reading *Psychology Today* and more reading *Guns and Ammo.* But indulge me in a conceit. Reflect on the fact that what we now call "the Old South" was actually a rather young region (some patches of Tidewater aside): gawky, awkward, sometimes rambunctious, testing the limits of its independence. After the region went to war and came back sadder and wiser, its concerns were those of young adulthood: making a living, principally. And now, after some decades of that effort have begun to pay off, the South, entering middle age, can finally afford to ask those potentially debilitating questions: What's it all about? Is this it? Who am I?

We need answers to those questions and, one way or another, we are going to get them. Clearly the old answers, the ones that were assumed while the South got about the business of making a living, will not do anymore. Sure, the South could cling to the old answers anyway, becoming in the process narrow, rigid, crabby, and socially isolated. Or, figuratively speaking, the region could get its hair styled, and start hanging out in singles bars. Even suicide would be an answer of sorts, but I believe that the South will survive. Southerners will continue to think of themselves as Southerners, although what that means will change—indeed, it has already begun to change. The new identity will incorporate some elements of the old; it will not be an entirely new construction. Some of its outlines are already apparent, but others are still up for grabs. One of the pleasures of

South-watching in the late twentieth century is speculating about how this story will turn out.

But I am getting ahead of myself. I wanted only to say that I am not using the phrase "mid-life crisis" in any rigorous or technical sense (I am not sure it can be used that way), just using it as a pretext for looking at regional identity. What are Southerners, anyway—or, more precisely, what do Southerners think they are? Talking about that question has been a favorite pastime of Southerners for almost as long as there has been a South, and some non-Southerners have been interested, too. "Tell about the South," said Quentin Compson's Canadian roommate at Harvard, provoking *Absalom, Absalom!* and providing Fred Hobson with a title for his recent book.

As often as not, those who have tried to tell about the South have sought some one defining characteristic, a single institution or peculiarity that has made the South what it is, a "central theme" in Southern history. Indeed, David Smiley wrote once that seeking a central theme in Southern history has been the central theme of the history of Southern history.

Many different answers have been suggested. David Potter, trying to convey just how puzzling the South is, called it "this sphinx on the American land." Ask it questions and you get answers, but figuring out what the answers mean is another matter altogether. If we must be classical about it, though, perhaps a better personification would be Proteus, that old man of the sea who could take any shape. Again and again, people have looked at the South and seized what they thought was its essence. Again and again, they have been left empty-handed when the thing changed shape. Yet the thing itself remained.

Naturally, antebellum Southerners were inclined to think that the South was about slavery. On the eve of secession, one South Carolinian wrote this: "The South is now in the formation of a Slave Republic [in which slavery] will make its stand, will build itself a home, and [will] erect for itself . . . a structure of imperial power and grandeur—a glorious Confederacy of States that will stand aloft and serene for ages." Five years later, that vision of the South was—well, it was gone with the wind. But the South was still there, more "solid" and perhaps more distinctive without slavery than it had been with it.

Sixty years later still, Ulrich B. Phillips, a distinguished historian of the South, said that white supremacy was what the South was about. A devotion to that principle, he wrote, was "the central theme of Southern history and the cardinal test of a Southerner." Many have agreed with him, but even as he wrote Jim Crow was terminally ill—or so it seems with hindsight. Phillips would not recognize the South today. Proteus has changed shape again.

A few years after that, sixty-odd years ago, the Agrarian authors of *I'll Take My Stand* identified the South with agriculture and rural life. They did it eloquently, but it seems that they were wrong. Who would make that identification today, now that two-thirds of us live in cities, towns, and suburbs?

Others have identified the South with one-party politics, one-crop agriculture, One-Way religion; with the legacy of Civil War and Reconstruction; with a Cavalier tradition from the English Civil War or, more plausibly, with a Celtic heritage from Great Britain's geographical fringes—and many of these factors have been undeniably important in making the South what it is. But it is probably a mistake to make any one of them the sine qua non of Southernness. One after another, these features have faded or been replaced, and their champions have written their epitaphs for Dixie. But Proteus is still kicking.

Many of these efforts to define the South have had a despairing tone, echoing John C. Calhoun's deathbed words: "The South! The South! God knows what will become of her!" The South was virtually identified with the problems of which it certainly had its share. As Erskine Caldwell summed it up: "The South has always been shoved around like a country cousin. It buys mill-ends and wears hand-me-downs. It sits at second-table and is fed short-rations. . . . It is that dogtown on the other side of the railroad tracks that smells so badly [*sic*] every time the wind changes. . . . The South has been taking a beating for a long time."

In the 1940s, I am told, a schoolbook called *Contemporary Georgia* drew such a sorry picture of that state that a generation of schoolchildren knew it as "Contemptible Georgia." When Yankees were doing the defining they often took the same tack: the South was where the problems were. H. L. Mencken called it the "Sahara of the Bozart," a cultural wasteland populated by cretins whose long stretches of indolence were interrupted only by episodes of violence against animals, or each other.

Alas, Southerners have often responded with defenses that did not help much, replying in the terms laid down by our critics and bragging about having the first, the biggest, the newest, the fastest-growing . . . whatever. In his essay, Mencken compiled some of the feeble responses to an earlier article "deploring the arrested cultural development of Georgia," among them:

> The first to suggest putting to music Heber's "From Greenland's Icy Mountains" was Mrs. F. R. Goulding, of Savannah.
>
> Who does not recall with pleasure the writings of . . . Frank L. Stanton, Georgia's brilliant poet?
>
> Georgia was the first state to organize a Boys' Corn Club in the South—Newton county, 1904.

And so forth. It all reminds me of nothing so much as a book published in Toronto in 1973 called *1001 Reasons to Be Proud of Being Canadian.* (Number 38 is: "Because Canada's shape is almost perfectly square.")

It is no accident that Mencken's correspondents were from Georgia. Georgians seem always to have taken the lead in the first/biggest/newest defense of the South. They may even outdo Texans. A hundred years after Atlanta's Henry Grady popularized the phrase *New South*—meaning a South as good as the North, in the North's own terms—I saw an advertisement in Atlanta for Stone Mountain that billed it as "the world's largest stone carving," without mentioning what it is a carving *of.*

Outside Georgia, some Southerners respond differently to criticism. "Come to Arkansas, Mr. Mencken," one newspaper in that state growled. "Come to Arkansas . . . and get your liver drained."

This sectional psychological warfare, this continuation of the War Between the States by other means, is not especially edifying, but it can help us understand what it means to be Southern. Any single answer is insufficient by itself, but notice that each of these answers relies on a comparison. The one constant is that Southerners are different from Northerners, with respect to whatever is being discussed.

Slavery takes on significance only when it exists side-by-side with free society; otherwise it is just part of the human condition. Racism is highlighted and seems a good deal odder when it is found in part of a society that claims to be committed to assimilating the huddled masses of the Eastern Hemisphere. As C. Vann Woodward pointed out, conquest and defeat, frustration and humiliation, poverty and hard times look to most of the world as just the way things are; it takes a victorious, prosperous, and innocent North to make those things appear to be "Southern" experiences.

Lurking in the background of almost any statement about the South, if not stated outright, is a comparison to the rest of the United States. If people say Southerners are lazy, or violent, or religious, they mean: compared to Northerners—not to Mexicans, to Martians, or to some absolute standard.

A few years ago, Glenn Elder and I asked some of our students to list "typical Southern traits," "typical Northern traits," and (later) "typical American traits." What we found was that the "Southern traits" were virtually the mirror image of the "Northern" ones: slow, not fast; generous, not greedy; religious, not materialistic; conservative, not progressive; and so forth. Leave aside whether these are accurate generalizations; the point is that when we asked these Southern students to tell about the South, they did it by comparison to the North. In the jargon of my trade, "Southerner" seems to be a reactive identity. It is almost as if we need Northerners around to know who we are.

Those students did another interesting thing. The list of "American" traits they gave us had almost nothing in common with their list of "Southern" traits; it was in fact very similar to their list of "Northern" ones. This does not mean that they did not think of themselves as Americans, just that their standard for comparison had changed. Compared to other Americans, they were Southerners: traditional, easygoing, polite, and so forth. But compared to foreigners, they were Americans—progressive, efficient, and all the rest.

Both Southerners and non-Southerners do think of Southerners primarily as Americans, but notice that there are other possibilities. My friend Edgar Thompson of Duke University, who knew as much about plantations as anyone I have ever known, found it useful to think of the South as the English-speaking, Protestant, northern end of a plantation region that once stretched down into the Caribbean and Latin America. But that is a useful way to think precisely because it is so strange.

Among us Americans, though, our differences are what define Southerners as Southerners. To the extent that W. J. Cash was right when he wrote that we can still speak of "one South" despite the region's diversity—to the extent that the South is held together—Southerners' belief that they have distinctive things in common supplies the glue.

Of course, the South has always been a messy assortment of landscapes, local societies, economic modes, peoples, and cultures. What do Georgia and Arkansas have in common, after all, or Kentucky and Louisiana? There certainly are a lot of differences and it is tempting to dwell on them, but bring Massachusetts into the picture, or California, and then it is a different story. South Carolinians will allow of Tennessee that at least it is not populated by Yankees, and South Carolinians will believe that means something.

What it means has certainly changed, though, and especially in the recent past. In his book *Journeys through the South,* Fred Powledge tells about what happened when an undergraduate geography class at the University of North Carolina in Charlotte tried to define the South. The students talked about accents, about word choices, about preferences in food, about the South's "slower pace"—but nobody mentioned slavery, or secession, or segregation. I tried it myself in Chapel Hill, with similar results. My students mentioned a more leisurely style of life, better manners, more conservative political and moral values, family pride and loyalty, a great many different things—but they produced a remarkably ahistorical list. Nobody mentioned that some white Southerners once owned slaves, or that they lost a war, or that they practiced segregation for some decades. Not even my black students mentioned these things, although that may just have been their Southern good manners. I was doubtful that this experiment would turn out the same in the Deep South, but when a friend tried it in Alabama, it did. (I am virtually certain that *Northern* students would mention these unhappy facts, but so far as I know no one has tried this exercise with them.)

What is going on? Are they teaching no-fault history in high schools these days? Are they teaching history at all?

Well, in fact, some educators have tried to promote a little healthy ignorance. The authorities who found Jim Loewen's Mississippi history textbook unsuitable for use in that state's high schools objected to a photograph of a lynching (among other things). One official protested that all that stuff is in the past—which Jim said he thought was a strange objection to a history book.

One student did ask me during a test to remind her which side was the Union and which the Confederacy, but she was unusual. I think most students know the basic facts of Southern history and acknowledge them readily enough when asked. It is just that history is not what comes to mind when they are asked what the South is about. The burden of Southern history—to borrow Professor Woodward's fine phrase—does not seem to weigh heavily on them; it has little to do with their sense of themselves as Southerners. Most seem to think about the Civil War about as often as the War of the Roses.

I am certain that is a change, and an important one. I am not so sure whether it is a good one or not. It may be that a happy historical amnesia will let us get on with our lives in this region, allow us to "forget the bad and keep the good" (as a country song by Tanya Tucker puts it). On the other hand, I am reminded of a conversation I once had with some young Germans. I suggested that we had something in common: How did it feel to them, I asked, to have lost a war, to recognize that the world may be a better place for it, but nevertheless to know that you lost? They simply did not understand the question. They seemed to have no sense that *they* had lost a war, no sense of identification with the generation that had happened to. For some reason, I found that disturbing.

In any case, survey research suggests that regional identification is still lively. For many in the South, "Southerner" is still a significant part of the answer to the question "Who are you?" But what Southerners now see themselves as having in common, whatever it may have been before, is less a tragic and un-American history (or, for that matter, a glorious and super-American history) than a shared culture in the present—that, and some shared complaints about the rest of the country.

People find this business of culture difficult to articulate, and I do, too, but Michael O'Brien, in his book *The Idea of the American South,* reminds us that the work of the political scientist Karl Deutsch is helpful. Deutsch is concerned with what makes a nation, and he equates "nation-ness" with ease of communication. A nation is a community within which people can communicate easily. As Deutsch points out, a common culture makes communication more efficient.

In the extreme case, part of that common culture is a language that nobody else understands. Since a common dialect or even an accent is a step in that direction, making it easier for insiders to understand one another and harder for them to communicate with outsiders, perhaps it is no accident that the Southern accent turns up near the top of many people's lists of what defines the South.

But other aspects of culture also facilitate communication. Many assumptions are programmed into us as we grow up; many things "go without saying." There are shared ground rules for interaction that are mutually understood and that do not have to be explained or negotiated.

Often these understandings are embodied in *manners,* another of the regional differences usually mentioned when people are talking about the South. Manners prescribe, for instance, what it is appropriate to know about the person you are dealing with. An article a while back in *U.S. News and World Report* on Northern migrants to the South mentioned the frequent "misunderstandings" that come up, and told of a Northern woman's irritation at being asked by a neighbor where her husband was. What seems to Southerners to be courteous concern for someone else's relatives (you are supposed to ask about them even if you do not care) can apparently be seen by Northerners as impolite, as nosiness (you should not ask even if you do care).

Closely related are shared ethical understandings, about what is good; shared religious understandings, about what is true; shared aesthetic understandings, about what is beautiful. Shared understandings determine what is in good taste, what is funny, even what is edible. In *An Asian Anthropologist in the South,* Choong Soon Kim tells about his first trip to an Atlanta supermarket, where he confusedly bought canned dog food. He observes that dog *meat* would have been all right.

This business of food is no laughing matter. Most national groups have their distinctive dishes that outsiders find revolting but that insiders like, or profess to like, at least on ceremonial occasions. Probably the best example is the Scots' haggis—oatmeal and liver in a sheep's stomach; Grady McWhiney tells me that *Reed* is 76 percent Celtic, but I find that idea more than 24 percent repellent.

If we are looking for the South's national dish, fried chicken and country ham are not repulsive enough. Chitterlings are probably too repulsive. Okra and the Moon Pie are possibilities, but if we must choose a single dish there is no question what it should be. Roy Blount, Jr., wrote a poem about it one time that concludes:

> True grits, more grits, fish, grits, and collards;
> Life is good where grits are swallered.

Shared understandings—about grits and other things—make communication easier, and these are the kinds of things people tend to mention these days when you ask them what makes the South what it is.

Notice that whether people understand one another is a different question from whether they like each other. That is obviously related, but its answer depends on a great many other things. In particular, it may depend on whether their interests are the same or opposed—that is, on whether what is good for one is good or bad for the other.

But many Southerners do seem to believe that they have common interests. Our survey research tells us that some believe that whatever is good for the South's economy, whatever increases the South's political power, whatever improves its image is good for them personally. This perception of shared interest is the second of the bases for regional identification that seem to be persisting.

Actually, these beliefs are most often expressed as grievances: people tell us that, as Southerners, they are economically exploited, politically discriminated against, or—most common these days—denied the respect due them as Americans, or human beings. Notice, though, that these are grievances about how Southerners are treated *in the present*. I do not deny that those Southerners with a bone-deep sense of history will have a regional identity shaped and nourished by that history, and the results can be wonderful indeed, as Southern literature continues to attest. But the sense of being Southern seldom grows from that soil, these days. When we ask what it means to be Southern our respondents talk instead about what Southerners have in common *now:* on the one hand, aspects of language and culture that make it easier for Southerners to understand one another; on the other hand, a sense of shared ill-treatment at the hands of the rest of the country—in particular, a sense of being looked down on by other Americans.

In a book called *Southerners: The Social Psychology of Sectionalism,* I looked at how modern conditions—urban life, expanded education, the mass media, more travel, and so forth—can actually heighten regional identification by producing greater awareness of Southern "things in common." Here, let me just point out a couple of implications of a quasi-ethnic identity that is based less on a shared history and the ancestry that transmits that history to the present than on a common cultural style and a shared set of grievances.

One implication is that a group defined by this sort of identification can be extraordinarily open; its boundaries can be remarkably permeable. Anyone who adopts the cultural style and voices the appropriate grievances is a potential member. This is one of the weakest points in the argument some of us have made that Southerners can be regarded as an ethnic group.

Ethnic groups are defined by a belief in a common ancestry, a set of shared experiences that happened in the past, to members' forebears, not to them directly. This puts limits on how open ethnic groups can be to outsiders. Maybe there is some sense in which I could become, say, an Italian-American, against all the known facts of my genealogy, but I would have to work at it pretty hard, and that sort of transformation cannot be expected to happen very often. As Horace Kallen put it, in a famous essay on the "melting pot": "Men may change their clothes, their politics, their wives, their religions, their philosophies, to a greater or lesser extent; they cannot change their grandfathers."

But if I am right about how Southern identification works these days, to become Southerners outsiders don't have to change their grandfathers. They don't even have to denounce them. If the migrants in the current wave want to become Southerners, they can probably do it. Whether they want it or not, many of their children will probably succumb. All they have to do is talk and act like Southerners. (Part of talking and acting like a Southerner, of course, is complaining about how the South is treated.)

I know a Boston-Irish family who now live in Winston-Salem. The youngest of the five children was a student of mine. His mother hates Winston-Salem, hates North Carolina, hates the South—and has ever since her husband got a job here twenty-five years ago. She goes back to Boston every chance she gets and is looking forward to moving back there when her husband retires. But right in the bosom of her family is a boy, my student, who looks and sounds and seemingly thinks very much like Bo, of "The Dukes of Hazzard." He used to put his boots up on the trash can in my office and complain about his mother the Yankee and her unreasonable dislike for the South. He is staying.

The South, as George Tindall has observed, has always been able to absorb outsiders. "Over the years," he wrote, "all those Southerners with names like Kruttschnitt, Kolb, DeBardeleben, Huger, Lanneau, Toledano, Moise, Jastremski, or Cheros got melted down and poured back out in the mold of good old boys and girls, if not of the gentry. Who, for example, could be more WASPish than Scarlett O'Hara, in more ways than one?"

When Choong Soon Kim studied the Choctaws of Mississippi, he observed that even they had been thoroughly assimilated to rural Southern ways; a visiting Plains Indian noticed the same thing, and grumbled that the Choctaws were not redskins at all, just rednecks. My favorite example of assimilation, though, is a country music disk jockey I once heard on a Georgia radio station. His name—I swear it—was John Wesley Cohen.

The big exception to this happy cultural gumbo, of course, has been Southern blacks, who have not by and large been inclined or encouraged to think of themselves as black Southerners. But I do believe this is changing. On William Buckley's "Firing Line" in 1982, William Ferris, director of the Center for the Study of Southern Culture at Ole Miss, told a somewhat skeptical Buckley that "in the decade of the '80s what we're seeing is an interesting kind of evolution from the '60s and the '70s to a sense of Southerners as Southerners as opposed to black versus white." Maybe so. If I am right about how Southern identification works, certainly the possibility is there.

The similarities of style and culture have always been there, of course; it could hardly be otherwise. In the past almost everyone, black and white, seemed to find the differences more important, but maybe the time has come to emphasize the similarities. Certainly those similarities seem to be especially evident to black and white Southerners who have come to know one another outside the South, and that experience is increasingly common. It may have had something to do with what Bill Ferris told Bill Buckley: Ferris, after all, came back to Mississippi from the Afro-American Studies Program at Yale. Not to belabor the point, but anyone who needs convincing should read Albert Murray's splendid *South to a Very Old Place,* a book with the bad luck to be published at least ten years before its time.

A while back, Merle Black and I reported some data that showed a striking increase between 1964 and 1976 in the tendency of Southern blacks to think of themselves as Southerners. We will know the process is complete when black Southerners habitually complain that non-Southerners do not understand the South, that they are tired of hearing the South put down, and so on. It has already started to happen. Robert Botsch turned up a nice example when a black North Carolina furniture worker told him he was planning to vote for Jimmy Carter because he was "tired of listening to all those slick Yankees who think they know everything and have all the answers." It is hard to sound more Southern than that.

Let me wrap this up by summarizing. I am saying that if you want to know what Southerners are, you could do worse than to ask them what they think they are, and, for most, the answer these days has little to do with ancestry, with the Civil War, with the Peculiar Institution or any of the South's other peculiar institutions. Instead, what we hear is that Southerners share some things in the present: first, what one anthropologist has called an ethnic style (not the same as an ethnic tradition); second, some mistreatment at the hands, or mouths, of the rest of the country.

And that has some implications that I, at least, find cheering. It means that migrants to the South, or their children, will probably prove to be assimilable, which is good news for those of us who like to have the South around. In addition, it opens up at least the possibility of a regional community that would cross the racial divide. It would be ironic indeed, but wholly delightful, if a sense of regional identification that has been bound up in the past with the struggle to maintain white supremacy could serve to bind up the wounds that struggle produced. We could surprise the cynics yet.

New South or No South?

Southern Culture in 2036

■ The year 1986 marked several significant Southern anniversaries. It had been fifty years since the founding of the Southern Sociological Society; fifty years since the publication of *Gone with the Wind;* fifty years since Howard W. Odum's classic *Southern Regions of the United States;* and, not least, an even century since Henry Grady, editor of the *Atlanta Constitution,* gave a speech in Boston that popularized the phrase *New South.*

There is a sort of numerological magic about the round numbers, these half-centuries and centuries (not to mention millennia, as we will have occasion to observe soon enough), and we probably ought to remind ourselves that these divisions are social conventions, reflecting only the evolutionary accident or divine whim that gave us ten fingers. Still, it has been a hundred years since Grady's speech, which does mean that very few who were living then are still with us; certainly nobody remembers the speech firsthand.

But, since 1886, many of us have heard a great deal about its subject. As Edwin Yoder put it, different versions of the New South have come and gone "like French constitutions and theories of the decline of Rome." Talk of a New South probably reached its high point in 1976, when Americans elected the first undeniably Southern president since Andrew Johnson. So much was said about it then that Walker Percy was driven to complain:

> Of all the things I'm fed up with, I think I'm fed up most with hearing about the New South. . . . One of the first things I can remember in my life was hearing about the New South. I was three years old, in Alabama. Not a year has passed since that I haven't

> heard about a new South. I would dearly love never to hear the New South mentioned again. In fact my definition of a new South would be a South in which it never occurred to anybody to mention the New South.

Percy is right, of course. But the phrase is convenient shorthand for whatever the South is becoming. And I want to ask a question about the next "New South": how much and in what respects will its culture be assimilated to national patterns?

Continuing Southern Distinctiveness

There was a time when few even thought to ask that question about Southern culture. Nearly everyone just assumed that the South would remain culturally distinctive. Certainly Henry Grady did. In his "New South" speech, Grady made some concession to the fact that the North had won the Civil War; he was willing, even eager, to adopt Northeastern economic ways, to build an urban, industrial economy like the one that had defeated his own. But scholars like Paul Gaston have pointed out quite rightly that Grady and his many admirers intended to pour the old cultural wine of planter rule and white supremacy into those new economic and demographic bottles. It apparently never crossed their minds to wonder whether traditional Southern ways were consistent with life in a radically altered society.

Later, though, particularly as Grady's economic prescriptions actually began to take effect, others began to wonder. As Broadus Mitchell put it in 1928, will "these great industrial developments . . . banish the personality of the South," or will industrialism "submit to be modified by a persistent Southern temperament?" Scholars from many disciplines—historians and anthropologists, folklorists and students of literature, geographers and political scientists—all have applied their disciplines' characteristic modes of inquiry to this question. Many brilliant and sensitive journalists have asked that question, too, and it is foolishness, mere academic snobbery, to ignore what they have written. Their work can be called "impressionistic" by those who wish to dismiss it, but at its best it is brilliant and sensitive ethnography.

One of the greatest of those journalists, W. J. Cash, put the case for cultural persistence as strongly as it has ever been put, in his masterpiece, *The Mind of the South.* Cash thought even Southern skyscrapers were expressions of continuity, reflecting an old tradition of civic pride and the search for glory. "Softly," he asked, "do you not hear behind that the gallop of Jeb Stuart's cavalrymen?" Not everyone has agreed, of course—C. Vann Woodward replied to Cash's rhetorical question: "The answer is 'no'! Not one ghostly echo of a gallop"—and that argument itself is on the way to becoming a persistent feature of our region's culture.

Few these days deny that the South has always been changing, and will continue to change as it moves into the next century. The disagreement today is about the extent and nature of the changes. The question has become: Does "New South" mean no South?

The Sociological Contribution

Oddly enough, it is only recently that sociologists have really had anything to say about this, despite the fact that it would seem on the face of it to be a sociological question. (The "regional sociologists" of the thirties and forties had other concerns, pressing ones related to the South's painfully evident social problems.) But the past quarter-century has seen the emergence of what is now a fairly large body of sociological literature, most of it based on survey research into attitudes and values. Lewis Killian, Andrew Greeley, Norval Glenn and his colleagues, Harry Holloway and Ted Robinson, Jeanne Hurlbert, myself—all of us have addressed one or more articles or books to the question. For the most part, these studies have adopted the rough-and-ready expedient of defining "Southerners" as residents of the Southern states (themselves variously defined) and comparing these respondents to other Americans on selected measures of attitudes, values, and behavior. Some studies have examined trends in regional difference over time; others have compared regional differences among younger Americans to those found among older ones; most have introduced controls for such confounding factors as education, occupation, and size of place. The methodological shortcomings of these studies are obvious enough, I should think, but they may be offset by the fact that the question of convergence is by its nature quantitative; survey methods can address it with a precision no other approach offers, a precision that may be misleading but one that at least establishes which side of the argument must bear the burden of proof.

In any case, this accumulating literature demonstrates that some important regional differences in the United States are getting smaller: many of them because the South is coming to resemble the rest of the United States; a few because the rest of the United States is starting to look like the South. But, at the same time, it shows that a great many large regional differences still exist. Some of these differences are not going away, and some, indeed, are increasing. In other words, nearly all the logical possibilities of what could be happening *are* happening, in one respect or another.

Can anything sensible be said about which differences are doing what? I think we can begin to make sense of this mass of data by distinguishing between two different sorts of attitudes and values that have, until recently, been "Southern" ones—by which I mean only that they have been found more often among Southerners than among other Americans.

Disappearing Traditional Values

In the first place, Southerners in the aggregate, black and white, have been characterized by a constellation of values that are common to folk, village, and peasant cultures everywhere in the world—not surprisingly, since that is what the whole South was as recently as fifty years ago, and what parts of it still are today. Any student of modernization could list these attributes; it seems perfectly clear that they are linked to economic development, and we do not have to take sides here in the quarrel about whether they are cause or effect. Among these attributes would be such things as fatalism and suspicion of innovation. Other traditional values seem to involve emphasizing firm and fixed boundaries between categories: between family members and others, for instance; between local people and strangers, or between ethnic brethren and aliens; between men and women, leaders and followers, and so forth. Indeed, cultural modernization seems to consist largely of learning to live with fuzzy and ambiguous boundaries.

Harold Grasmick, whose literature review and research on this subject I have leaned on heavily, calls this cluster of values "the" traditional value orientation, and he shows that, among Southerners, it is eroded by urban life, by education, by travel and residence outside the South, by exposure to the mass media. Since all these experiences are more common in each new generation of Southerners, it is not surprising to find that the traditional value orientation is less common among young Southerners than old ones, and less common now than a generation ago.

One consequence of the continuing economic and demographic convergence between the South and the rest of the United States will surely be regional convergence with respect to these traditional values as well. Indeed, the process has been going on slowly but steadily for some time now, and there is no reason to suppose that convergence will not soon be complete. Already regional differences in religious prejudice have virtually vanished, and differences in racial prejudice are so much smaller now than twenty years ago that they can almost be ignored. (Some of the demographic differences, however, have disappeared only to reemerge as differences in the opposite direction, as Ronald Rindfuss has shown, and there is always the possibility that the same may happen with some of the decreasing cultural differences.)

Another factor to consider, of course, is migration to the South. If it continues at anything like its present level, it will also nudge regional convergence along, because migrants to the South are conspicuously "untraditional": they are, that is, much less likely to display these characteristics than native Southerners, white or (especially) black. (By most of these measures, incidentally, blacks are the most "Southern" of all Southerners.)

Enduring Differences

But there have been other regional differences. Some Southern characteristics are not linked in any obvious way to rural residence, agricultural pursuits, poor education and not much of it, limited exposure to the "wider world." In *The Enduring South,* I examined some of them and ventured to predict that regional differences in some of these characteristics are going to be with us for a while.

Although it is presumptuous for me to mention my book in the same breath as Cash's *Mind of the South,* there is at least a genetic connection between the two books that I would like to put on the record. When I first read *The Mind of the South,* for recreation, in graduate school, I found it methodologically exasperating. I kept asking (as I was being trained to ask), "How does he know that?" When it came time to write a dissertation, I set out to study the mind of the South the way my teacher Paul Lazarsfeld had studied "the academic mind"—that is, with survey data. The result became *The Enduring South.* And I must say I emerged from the experience with a new respect for the achievement, and the "methodology," of W. J. Cash.

The differences my book documented would not have surprised Cash at all. Most of them can be seen as manifestations of the individualism that Cash saw as central to Southern culture. They reflect an anti-institutional ethic that says: In the last analysis, you are on your own—and should be. Southerners in 1900, Cash wrote, "would see the world in much the same terms in which their fathers had seen it in 1830; as, in its last aspect, a simple solution, an aggregation of self-contained and self-sufficient monads, each of whom was ultimately and completely responsible for himself."

Well, Walter Hines Page said once that "next to fried foods, the South has suffered most from oratory," and maybe Cash got carried away. But certainly an individualistic, anti-institutional note is evident in many aspects of Southern life. It is obvious, for example, in the Evangelical Protestantism to which most Southerners subscribe, a form of religion that sees salvation as something to be worked out by the individual, in a direct, unmediated relationship with Jesus. As Tom T. Hall sings it, in a classic country song, he and Jesus have their own thing going, and don't need anybody to tell them what it's all about.

The same pattern is evident, I think, in a disposition to redress grievances privately, which sometimes means violently—a disposition that gives black and white Southerners homicide rates at which the rest of the civilized world marvels. Anti-institutionalism can also be seen in a sort of localism and familism that reflects, not ignorance of the alternatives (as in the traditional value orientation), but simple preference for the palpable and close at hand, as opposed to the distant and formal. And, finally, individualism may be reflected in a sort of economic libertarianism that was apparently suppressed during the hard times of the past 120 years, but that seems to be coming back strong in our own times, at least among Southern whites.

All of these are "Southern" attributes, more common in the South than elsewhere. And in these respects, the South is not becoming more like the rest of the country: indeed, in some of them, regional differences are increasing. Unlike the traditional value orientation, these attributes are by no means universal among "folk societies," nor are they limited to such societies. And their statistical behavior is different, too: some of these traits are more common among educated Southerners than uneducated ones, or among urban Southerners than rural ones. Economic conservatism, as Robert Freymeyer has shown, is even more common among migrants to the South than among natives, so one cannot expect migration to reduce that difference.

Origins of Southern Anti-Institutionalism

Where these traits came from in the first place is an interesting question, but it is obviously a historical one, and it is tempting to leave it to the historians, especially since they disagree among themselves. Some point to the ethnic origins of Southerners and argue a Celtic influence, or an African one, or some catalytic interaction of the two—each an explanation that at least moves the problem back across the Atlantic and out of the hair of American historians. Others point to the legacy of the plantation, or the persistence of the frontier, or one of a half-dozen other "central themes" of Southern history. A sociologist would be foolhardy indeed to step into that fracas, but I will venture to suspect that the explanation for continuing Southern distinctiveness does repose somehow in the fact that black and white Southerners are *groups to which things have happened,* groups whose members have learned lessons from their collective histories.

In one well-known version of this argument, C. Vann Woodward has suggested that the un-American experiences of defeat, military occupation, poverty, frustration, and moral guilt have given white Southerners an outlook qualitatively different from that of other Americans. Obviously, a similar argument could easily be made for black Southerners (leave aside the moral guilt). This has been a very appealing line of thought, and there may be something to it, although it leaves a number of questions unanswered. This is not the place for a critique of Woodward's argument, much less a test of his hypothesis, but let me just observe that he does not spell out how events and experiences of a century or more ago are supposed to affect values and attitudes today. My own research suggests that it is seldom a matter of lessons continuing to be learned from reflections on that history: indeed, artists and historians aside, few Southerners of either race seem to know or to care much about their groups' histories. Rather, it appears that lessons drawn from group experience may have been passed on to succeeding generations without knowledge of the facts from which the conclusions were drawn.

It is not clear either exactly what lessons Southern history is supposed to have taught. Returning to our pattern of individualism and anti-institutionalism, however, it seems to me that long experience with an environment seemingly uncaring or even hostile could produce or reinforce traits like those to be explained. And both black and white Southerners, in their different ways, have had good reason in the past to view their environments as unresponsive if not actually malevolent.

The Future of Regional Differences

What difference does it make where this pattern of individualism comes from? Well, it has some implications for what we can expect as we move into the next century. Regional differences brought about by different historical experiences will not go away simply because Southerners come to have economic and demographic circumstances like those of other Americans. This sort of "Southern characteristic" could prove much more durable—apparently is proving much more durable—than traits that reflect simply economic underdevelopment.

Some might want to argue that group differences tend to decrease in the absence of forces operating to maintain them—sort of a cultural version of the Second Law of Thermodynamics—but that is not at all a self-evident proposition; at least its validity needs to be demonstrated.

All this is speculation, of course, but if this view is correct, regional differences of this sort will decrease if history teaches its lessons more indiscriminately. Some course of events might cause Southerners to forget the lessons of their past, or teach other Americans what have heretofore been "Southern" lessons. Either way, regional differences would decrease.

In 1986, the revised edition of *The Enduring South* brought many of the data in the book forward another twenty years, from about the mid-1960s to the 1980s. For the most part, the book's original analysis held up pretty well. The regional differences that should have become smaller did so. Some cultural differences were largely due to Southerners' lower incomes and educational levels, to their predominantly rural and small-town residence, to their concentration in agricultural and low-level industrial occupations. Those differences were smaller in the 1960s than they had been in the past, and they are smaller still today. A few, indeed, have vanished altogether. There are important respects in which Southerners look more like other Americans, culturally, than they have at any time for decades, if ever.

On the other hand, differences that were persisting in the 1960s—in localism (as *The Enduring South* measured it), in attitudes toward some sorts of violence, in a number of religious and quasi-religious beliefs and behaviors—are mostly still with us, usually as large as they were then, and occasionally larger. This, I believe, supports the book's conclusion that those are differences of a quasi-ethnic sort, with their origins in the different histories of American regional groups, not merely epiphenomena of different levels of current economic and demographic "modernization."

There were, however, a few—only a very few—instances where these differences, too, were smaller than a generation before. And where that happened it was not because the South became less Southern, but because the non-South changed to resemble the South.

This pattern is evident in survey items of several different sorts, but my favorite example involves responses to the question, "What man that you have heard or read about, living today in any part of the world, do you admire the most?" When the Gallup Poll asks this question, as it often does, it is obviously fishing for the name of a public figure. Nevertheless, many Southerners perversely insist on naming people the interviewer never heard of: relatives, friends, miscellaneous local figures. In 1965, nearly a third of white Southern respondents did that, and they were almost twice as likely as non-Southerners to do so.

If this response indicates a species of localism, by this measure the regional difference in localism had decreased by 1980: from 14 percentage points to 5. But not because Southerners had become less localistic. The percentage of Southerners who refused to name a public figure had actually risen, from 32 percent to 38 percent, but the percentage of non-Southerners who gave this kind of response had risen, too, even faster: from 18 percent to 33 percent. (The same pattern can be observed, by the way, with a similar question about most-admired women.)

This is not the model of regional convergence that most observers have had in mind. But the 1960s and 1970s gave all Americans some un-American experiences: Vietnam, Watergate, Iran, assassinations, urban riots, economic stagnation, double-digit inflation, rising crime rates, urban decay, and a host of other distressing, frustrating, alienating developments. Is it stretching a point to speculate that many non-Southerners have begun to believe, as many Southerners already did, that they cannot always get what they want or even keep what they have, that hard work and good intentions are not always enough, that politicians and bureaucracies are not to be trusted, that people out there don't like them? And if non-Southerners come to believe that, would it be surprising if they took up the values of a nearby subculture that came to those conclusions some time ago—for example, the value that says your family, friends, and neighbors are more reliably admirable than people you have only "heard or read about"?

Let me close with a general point. It is fairly easy to predict the future of regional cultural differences that are merely reflections of economic and demographic differences. All it takes are economic and demographic projections (and the willingness to trust them, which is the hard part). But if we allow that group culture can also be shaped by history, by events, prediction becomes much more difficult. It becomes nothing less than a matter of predicting events. If sociologists could do that any better than anyone else, we could put a lot of stockbrokers and fortune-tellers out of work.

Preserving the South's Quality of Life

■ Southerners like the South, warts and all. And many deny that there are any warts.

Item. The theme of nostalgia and affection for the South, conspicuous in American popular music during the first half of the twentieth century, survives—indeed, flourishes—in dozens of country music songs today: more so even than a generation ago.

Item. When the Gallup Poll asks Americans where they would live if they could live anywhere they wanted, Southerners are more likely than other Americans (save only Californians) to say "right here," and they have been more likely to say this for at least a half-century.

Item. Residents of Southern states are more likely than other Americans (again, except for Californians) to tell pollsters that the "best American state" is their own.

Item. Historically, Southerners have been at least as likely as other Americans to leave their home region, but the South's economic development of the past few decades appears to be changing that: fewer now leave, and many who left earlier are returning.

Item. Asked to name the worst thing about the South, a substantial minority of Southerners—one in five, in a North Carolina survey in the early 1970s—reply that there is nothing "worst" about the South.

Southerners who do dislike something about the South dislike the obvious. Who could celebrate the South's low wages, its industry mix, its poverty or unemployment rate, or its standing with regard to most of the things that governments routinely collect statistics about?

Those things aside, however, most Southerners seem to like what they have. As one North Carolinian told an interviewer: "I can't think of anything bad about [the South], other than we don't make too much money down here." Money is important, but it is not the only factor with consequences for a region's quality of life. Few Southerners use that phrase—they are more likely to talk about "the Southern way of life" or, regrettably, the Southern "life-style"—but most apparently feel that the South's noneconomic amenities offset any economic disadvantages when it comes to deciding what to call the "best American state," where to write songs about, where to live.

What do Southerners like about the South? What is this "way of life" that they are so fond of? Wherever we turn for an answer to that question—to the lyrics of country music, to the letters columns of Southern newspapers, to the archives of public opinion poll data—we get a remarkably consistent set of answers. If I concentrate on the last of these sources, it is only because the survey data are not so generally accessible as the others. Before I summarize those data, however, a couple of warnings are in order.

First, asking what people like about something—about the South, in this case—is an inherently conservative approach. Southerners cannot like aspects of Southern life that do not exist; we do not hear, for example, that the best thing about the South is its many ethnic restaurants, or opera companies. This does not mean that Southerners would not like these things if they existed, but we are asking—and so we are told—about what they like now. Those things may be worth preserving, but they are not the only good things imaginable.

In addition, we should bear in mind that when Southerners tell us about the South, they almost invariably do it with reference to the non-Southern United States. "What do you like about the South?" is almost always understood to imply "compared to the North." In consequence, Southerners simply do not mention important characteristics of Southern life, components of any sensible "quality-of-life" measure, if they are not seen as ways in which the South is different from the non-South. It goes without saying, apparently, that Southerners like about the South what they like about America in general. It certainly does not follow that these American "good things"—political freedom, the rule of law—are unimportant.

With those general limitations in mind, I can report that most Southerners apparently do believe that the South is different from the rest of the United States in important respects, and most prefer it that way. Individual Southerners like many different things about the South, some of them contradictory, but most agree about a couple of ways in which it is superior. When asked what they like about the South, majorities of Southerners mention pleasant natural conditions and pleasant people.

Natural Conditions

In "Dixie on My Mind," Hank Williams, Jr., sings about missing, more than anything else, his native South's rivers and pines. It seems that many Southerners share Bocephus's tastes, as indicated by this excerpt from a summary of the North Carolina survey reported in my book *Southerners*:

> Although some [respondents] complained about heat, humidity, insects, and the like when asked for "the worst thing about the South," most . . . seemed to like the physical environment of their region. When we asked what the best thing is, two-thirds mentioned the South's climate, its forests, mountains, or coast, its lack of crowding and pollution, the opportunities it offers for outdoor recreation. In general, they seemed to agree with William Faulkner's view that the South is fortunate that God has done so much for it, and man so little.

The report goes on to quote some of the survey respondents:

> [I like] the open spaces, the uncrowded roads and countryside. The physical environment agrees with me.
>
> The climate, the mild winter and change of seasons; the fact that air pollution and overcrowding isn't as prevalent as in the Northeast.
>
> It's green, clean-looking, not eat-up with pollution.
>
> I like the climate and open spaces and fresh food.
>
> It's beautiful here. A good place to bring children up. Living conditions are better in general. Recreation is better. I like to fish.

> I like the mountains, beaches. . . . This is just a good place to live. Not as crowded or polluted as the North.

> Used to be able to say clean air, but you can't say that any more. There's plenty of room to stretch out.

The report noted that this last theme—"room to stretch out"—was a common one: "[Many] volunteered that the South is 'not as crowded' as the North, that it still offers 'roominess' and 'wide open spaces,' that people are 'not piled up so bad.' Several mentioned 'freedom' in the same breath as lack of crowding." A good many respondents volunteered that the absence of big cities and big-city problems was an asset; almost no one saw the relative absence of cities as a problem per se.

Of course, these aspects of the South—seen and valued, ironically, as contributing to the region's quality of life—are by-products of economic underdevelopment, and they are now increasingly menaced by urban and suburban sprawl, by industrialization, by population increase. A further irony is that the public policies that could preserve some of these amenities often conflict with other Southern values, in particular with a widespread, absolutist view of property rights.

Whether one views the prickly Southern individualism that W. J. Cash wrote about in *The Mind of the South* as an obstacle or a resource, it is still with us. Roy Reed described it (exaggerating a bit, no doubt) for the readers of the *New York Times,* in 1976. The example he used to illustrate it is very much to the present point: "It is no accident that the most determined holdouts against land-use legislation in the United States are country people from the South. They will take care of their own land, and let the next man take care of his. If the next man puts in a rendering plant or a junkyard, that is his business." "For all the American encroachments," he continued (and I cannot resist quoting him),

> the South is inhabited and given [its] dominant tone by men—and women who acquiesce in this matter—who carry in their hearts or genes or livers or lights an ancient, God-credited belief that a man has a right to do as he pleases. A right to be let alone in whatever plain of triumph he has staked out and won for his own. A right to go to hell or climb to the stars or sit still and do nothing, just as he

> damn well pleases, without restraint from anybody else and most assuredly without interference from any government anywhere.

Reed believes that this orientation is doomed, that the fight was lost "about the time [a man] lost the irretrievable right to take a leak off his own front porch." But, he observes, "they have not yet taken [Southerners'] right to curse and defy"—as many would-be land-use planners can no doubt attest.

No, any governmental attempt to preserve for Southerners what they say they like about the Southern landscape will come in for a good deal of suspicion, especially if government tries to forbid certain activities. People—perhaps especially Southerners—do not like being told what they cannot do, especially with their own property.

On the other hand, I suspect much could be done by government encouragement of various other activities—those of private organizations like the Nature Conservancy, for example. If the subject is approached imaginatively and with sensitivity to what it is that Southerners actually value, Southern business and industry may even have an important role to play. In a study of the business elite of North Carolina, Paul Luebke found surprising support for at least some forms of wilderness and wildlife preservation. That it came from businessmen who like to hunt and fish suggests a powerful constituency willing to support conservation, so long as it is not called "environmentalism." This enthusiasm is not limited to wealthy duck hunters and dry-fly fishermen either, and for the time being at least it persists among urban Southerners: one survey showed Southern city folk more likely to hunt than rural people elsewhere in the United States.

Organizations like Ducks Unlimited, Bassmasters, and local hunting, fishing, and garden clubs may have a great deal to contribute. Certainly state and local governments would be foolish to ignore the commitment of these groups and the political and economic resources at their disposal.

Good People

Southerners see the South as different from the rest of the country not only in its natural resources, but in human terms as well: for starters (to quote Hank Williams, Jr., again), the South is a place where people say grace and people say "ma'am." In general, Southerners see themselves as slower, more conservative, and more polite and friendly than other Americans. Some studies have shown that other Americans are generally prepared to grant them those differences, although not all Americans are ready to allow that these comparisons are to the South's advantage. Not even all Southerners are.

But most Southerners, apparently, like what they see as "Southern" characteristics. When asked what they liked best about the South, half of the respondents to the survey mentioned earlier said the people. When asked to describe Southerners, most were unstinting in their praise:

> Far and away the most frequent characterization of Southerners (by these Southerners) was an elaboration on the theme that Southerners are good people, as several said in so many words. "Good," to our respondents, meant primarily pleasant to be around: "considerate," "friendly," "hospitable," "polite," "gentle," "gracious," "cordial," "genteel," "courteous," "congenial," "nice"—all of these our respondents' words, many of them used often.

Again, the report provides examples of typical respondents' remarks:

> People in the South for the most part are more cordial and courteous than those I've found in the North. People have more consideration of one another.
>
> It's a sort of graciousness—gentility and a hospitality feeling—in people native to the South.
>
> Well, I find people friendly here everywhere I go.
>
> I have a great feeling of being respected and welcomed here.
>
> [The best thing about the South] is the attitude and friendly people as a whole. We live more like God intended us to live in relation to one another.

The report notes that "dissenting opinions were rare and were almost entirely voiced by migrants from the North," but even many migrants felt the same. As the wife of a chemical plant manager, a ten-year resident of North Carolina, put it: "I like the slow pace. Also the friendliness."

Many Southerners believe that the texture of daily life is more pleasant in the South, perhaps especially in casual, impersonal interactions like those in the course of shopping or business. There is obviously a danger of regional chauvinism here, but consider the possibility that our survey respondents might be correct. I think they are, as a matter of fact. This sort of thing is difficult to measure; it does not appear in government statistics. But one recent University of Michigan survey shows higher percentages of Southerners than of non-Southerners reporting that their lives are "very friendly." Call this "pseudo-gemeinschaft" if you are cynical, but any civility that persists in the marketplace is worth noting, celebrating, and—if possible—preserving.

How much can be preserved remains, to me at least, an open question. Although an easygoing and cordial mode of interaction is not a universal characteristic of rural and village cultures, it seems obvious to me that in the South it has been linked to small-town and rural life. Certainly it is inconsistent with the stereotypical urban life—anonymous, transient, and characterized by hurried interaction with a great many others. Civility "costs" more and "pays" less in the metropolis. If Southern cities have so far remained "Southern," in this sense, it may be only because so many of their inhabitants grew up and learned their manners, styles, and values in small towns and the Southern countryside. The crucial question is what the next generation of urban Southerners (the first generation to be a majority of the Southern population) will look like.

In addition, Southern metropolitan areas have been, along with specialized retirement communities, the principal destination for migrants from outside the South. Although it is easy to overstate the cultural impact of migration on the South as a whole, there is no question that many of our metropolitan areas have been profoundly shaped by these newcomers. And when it comes to "manners, styles, and values," there is often some question of who is going to adjust to whom.

Of course, as Herbert Gans reminds us in *The Urban Villagers,* metropolitan areas are not culturally uniform—perhaps no one will be "assimilated." And Gans also reminds us that urban neighborhoods are not even all culturally "urban." If cultural characteristics of the small-town and rural South are preserved in Southern cities, it will be because a critical mass of Southern city-dwellers have contrived to become "urban villagers," in stable, settled neighborhoods, with small, established institutions—churches, schools, and the like—in effect, re-creating the small-town South in an urban setting.

The Future

"If Heaven Ain't a Lot Like Dixie," the ever-quotable Hank, Jr., sings, he doesn't want to go. But the things that Southerners most often say that they like about the South are threatened by urbanization, by industrialization, by in-migration—by factors, in other words, that seem inextricably connected to the economic modernization that is solving many of the South's problems. The question is not whether these good things can be preserved at their present level—they will not be—but rather how much is amenable to preservation at an acceptable cost.

All the Southern states and many Southern communities have plainly opted for a major governmental role in their economic development efforts. Is there any part for government in ameliorating the less pleasant consequences of that development? I have already indicated some of the problems in preserving the natural beauty and open spaces of the South; when it comes to preserving amiability and good manners—well, I would not know where to start. The difficulties of attracting new industry are trivial, by comparison.

It is not my purpose and anyway I am not competent to suggest what Southern policymakers ought to do about these issues. Let me give just two examples, though, of the sorts of policy that affect these aspects of the Southern way of life, and that should not be considered without regard to those effects. (These examples may demonstrate the incompetence I just mentioned.)

Tax law has obvious implications for land use, and it can be used to encourage some uses, while discouraging (though not forbidding) others. Several states have already acted to give tax relief to farmers near cities, whose land has often risen drastically in value due to speculative pressure. This policy is often defended on the grounds of justice for farmers, but note that it also serves the interests of dove-hunters, as well as the aesthetic needs of those who simply like open space nearby.

Similarly, the neighborhood school may be pedagogically inefficient, and it may sometimes be an unlawful impediment to desegregation, but generations of immigrants have demonstrated that it is also a powerful institution for preserving cultural traditions and values in an urban environment that is hostile to them. If urban Southerners are to maintain the rural and small-town ways that many Southerners say they like, neighborhood schools can contribute to that preservation. Certainly they should not be bulldozed without thought to their social and cultural functions, as well as their academic ones.

I suspect that government's most constructive role will often be simply to stay out of the way, to avoid placing obstacles in the path of private natural and cultural conservation efforts—at best, perhaps, to remove a few obstacles. I have mentioned some of the nongovernmental resources that may help the South to avoid Northern mistakes in Southern settings (no less a danger for being a cliché). Let me close by mentioning a couple more.

In the first place, there is still in the South a level of devotion and commitment to local communities unparalleled elsewhere. In the past, this has often been expressed as mindless boosterism: in my Tennessee hometown, we used to brag about how high on the list of nuclear targets we were. But the local pride behind that sort of foolishness is sound. Perhaps it will eventually turn against the despoilers of the Southern countryside.

Our regional obsession with growth was perfectly understandable, even laudable, in a region as poor as ours once was, and it is still appropriate in those parts of the South that have been bypassed by economic development. But in much of the South, I think, the enthusiasm is beginning to pale. A Chapel Hill savings and loan recently advertised, "We are proud of this community and we want to see it grow." Well, some of the rest of us are proud of this community, too, but our civic pride tells us that if Chapel Hill grows much more it will no longer be something to be proud of.

Another resource the South has in abundance is voluntary associations of private citizens, nearly all of them available at least occasionally for public service—innumerable garden clubs, civic clubs, historical and preservation societies, as well as the hunting and fishing clubs I mentioned earlier. Most of all, though, there are the churches of the South, institutions of great, and increasing, wealth and power. Most are, by tradition, as suspicious of government as government (thanks to the Supreme Court's reading of the Constitution) must be of them. They have always worked, however, for what they have seen as community well-being. (That is, after all, what the anti-liquor campaign has been all about.) No policymaker can afford to ignore the churches' past and potential contributions to all aspects of the South's quality of life.

In general, the lesson here for policymakers does not come down to a list of proposals, to an "action agenda" to preserve the endangered noneconomic aspects of the Southern way of life. Rather, it is an appeal to them to be sensitive to these matters when dealing with other things—with property tax law, with school sitings and school closings, with governmental relations to voluntary organizations of various kinds, with economic development policy.

Everyone, surely, would like to "increase per capita income, reduce poverty, and reduce unemployment." All these are worthy goals. There is nothing wrong with pottage; in fact, it is just the thing for a hungry person, or region. But most Southerners recognize —and policymakers should, too—that man does not live by bread alone. Pottage comes at a cost, but maybe that cost is negotiable.

Life and Leisure in the New South

■ In 1949 the *American Scholar,* the magazine of Phi Beta Kappa, published an article by H. C. Brearley with the title "Are Southerners Really Lazy?" Brearley's answer was somewhere between "probably not" and "yes, but it's not their fault." And he added that "even if the South has more than its proportionate share of the lazy and the incompetent, this is no reason to condemn all Southerners indiscriminately." He meant well, but that title raises some interesting questions besides the one it meant to raise: like, why don't Southerners share the immunity from ethnic slurs that other groups enjoy in polite circles? (There are no circles more polite than the *American Scholar.*)

My righteous indignation is undercut, however, by the fact that I intend to ask the same question. I want to examine Southerners' reputation for laziness (and the fact that we've been both reproached and envied for it); then—just among ourselves—I want to ask whether that reputation is deserved. *Are* Southerners really lazy? Is it true what they say about Dixie?

Reputation first. When Brearley asked whether Southerners are really lazy, he could assume that his readers had at least heard rumors to that effect. "Are Midwesterners Really Lazy?" wouldn't have been nearly as good a title. Laziness, sometimes under other names, has been part of the stereotype of Southerners for as long as there have been Southerners. Indeed, it has been part of Southerners' view of themselves. The image of Southerners as people who take life easy is so widespread that it can be *used.*

It can be used, for instance, to sell things. I saw an advertisement once that illustrates the point. "Southerners were never unduly worried at missing the only riverboat that week," said the headline. A photograph showed a riverbank, from which the *Robert E. Lee* had evidently just departed, and a table spread with shrimp, lobsters, crabs, oysters, and (for some obscure reason) bananas. Featured prominently was a bottle of Southern Comfort. The text informed us that "the Southern States of America's philosophy was never rush or panic. They took life easy. So missing a riverboat was no great problem. Even if it meant a seven-day wait until the next one." The message clearly was that if you drink enough Southern Comfort, you can be as nonchalant as a Southerner, and you won't care about missing riverboats either.

The only thing at all unusual about this advertisement is that I saw it in the magazine of the *Sunday Times* of London. We may be lazy, but we're *world famous* for it.

As I said, this reputation goes back a long way. Robert Beverley was writing in 1705 about the "lethargy" and "slothful indolence" of his fellow Virginians, and twenty years later William Byrd said some similar hard things about North Carolinians. (He said they did not distinguish Sunday from any other day of the week, which would be an advantage if they were hardworking. But "they keep so many Sabbaths every week that their disregard of the seventh day has no manner of cruelty in it either to servants or cattle.") Shortly after the Revolution, Thomas Jefferson wrote to a French friend that Southerners were "indolent, generous, and hot-tempered." Note that indolence heads the list.

Examples could be multiplied endlessly, but let one more suffice for now. In *The Mind of the South,* W. J. Cash wrote that Southern poor white males "mainly passed [their lives] on their backsides in the shade of a tree, communing with their hounds and a jug of what, with a fine feeling for words, had been named 'bust head.' " Even their farming, Cash added, "was largely confined to a little lackadaisical digging—largely by the women and children."

This attribution of laziness knows no bounds of class or race. The lassitude of slaves—who, after all, had no very good reason to exert themselves—was celebrated to the point where it was cited both by critics of slavery, who said free labor would be more productive, and by defenders of the institution, who said that slaves were not driven so hard that abolitionists needed to be concerned about them. Their inefficiency may sometimes have masked a form of passive resistance, but it seems to have exasperated Northern observers as much as it did slaveholders. Frederick Law Olmsted was not the only visitor who thought that many Southern masters were themselves of too "careless, temporizing, *shiftless* disposition" to work their slaves at full capacity. Certainly the laziness of the planter class comes through in fictional portrayals of the antebellum period. David Bertelson's book *The Lazy South* examines several plantation novels in this light (disapprovingly).

The perception we are examining is not confined to the kind of people who write books, or to outsiders. When Glen Elder and I asked a group of Southern college students to pick "typical Southern traits" from a list of eighty-four adjectives, "lazy" and "pleasure-loving" tied for seventh place. The trait most commonly attributed to *Northerners* was "industrious," with "ambitious" and "efficient" not far behind. (When Gail Wood repeated the study at a college in Pennsylvania, she found the same perceptions among Northern students.) When we asked a general population sample in North Carolina to tell us the "best thing" about the South, a substantial minority mentioned the region's "slower pace": "Southerners are not in such a run all the time—not so much hurry," one respondent told us. "People will talk to you." Another said that the best thing is "everything not being so hectic." "A less frantic tempo," said another. "We don't push like they do," said a fourth. (The ambivalence I mentioned before came through when we asked what the worst thing is. "Lack of ambition," said one man. "We're all lazy.")

Regional stereotypes are probably harmless enough, as ethnic stereotypes go, but they could result in harmful behavior—that is, in discrimination. We asked our North Carolina respondents to imagine that they managed a factory that had to hire a scientist. A Northerner and a Southerner apply. Which would they hire, and why? Half, probably to their credit, said they could not decide. Of the rest, however, one in eight chose the Northerner, and many of those said they did it because he could do the job "better and faster"; "would apply himself more"; would be "more interested in getting ahead and doing a better job"; or "would get the job done while the Southerner is thinking about it."

To be sure, many more were prepared to discriminate against the Northern applicant, but nobody mentioned the Southerner's ambition or efficiency as a reason, except for one man who said the

Southerner would "work at a more natural pace and be more productive than an aggressive person."

So a great deal of testimony, expert and otherwise, over a great span of time, says that Southerners are lazy, or easygoing, or laid-back, or inefficient, that they lack ambition, or push, or drive. . . . There are a variety of ways to put it, favorable and derogatory, but they all express much the same perception. As a bumper sticker I once saw said it: "Southerners do it slower."

Can all these testimonials be wrong? Well, sure they can. We can find equally impressive testimony to support nearly any common ethnic slur, and once these stereotypes are established they are extraordinarily resistant to change. Before coming to unflattering conclusions about any group, we ought to insist on better evidence than "common knowledge."

But there is a strange shortage of better evidence. In a recent article, for instance, Norval Glenn and Charles Weaver wanted to examine how Southern workers feel about work. "One might think," they complained, "that the answer to that question could be found in the hundreds of publications by social scientists who have systematically studied attitudes toward work in the United States. Surprisingly, however, the large literature on work attitudes provides no answer, nor does it provide any systematic evidence at all." In the absence of better data, Glenn and Weaver could not really begin to say whether Southerners like work more or less than other Americans, much less why. Nor is there much in the way of systematic evidence about Southern attitudes toward leisure.

But we can ask what Southerners *do* with their leisure, and here the evidence is somewhat better. It comes from a variety of sources and adds up to a remarkably consistent picture. It appears that, by and large, Southerners simply do *less* with their leisure time than other Americans—or at least less of most things that social scientists and market researchers are interested in.

Consider, for instance, the data in a publication called the *Target Group Index*. Published every year for the use of advertising and marketing types, the *Index* is based on an enormous sample of Americans and relates all sorts of commercially relevant behaviors to all manner of other characteristics. If you want to know, for example, what fraction of Tequila Sunrise drinkers watch "As the World Turns," the *Target Group Index* will tell you. There is a notice attached about what sorts of legal penalties will strike anyone who uses its data in unauthorized ways, but surely the publishers would not mind my saying that Southerners, one of the "target groups" examined, show lower levels of most of these activities. There's an exception here and there (Southerners drink more buttermilk, as you might suppose), but in general Southerners are not as big a target, not as good a market, as their numbers would suggest.

In part, this is almost certainly because the average Southerner still has less money to spend. That may begin to explain, too, why Southerners show lower levels of magazine and newspaper exposure, or even of miles driven per year (a substitute for exposure to billboard advertising). On the other hand, there is no real regional difference anymore in television and radio ownership, yet Southerners, on the average, watch television and listen to the radio less than other Americans. Here again, there are exceptions: college football and basketball games, "The Dukes of Hazzard," "Dallas," and daytime television in general attract more Southerners (or fewer non-Southerners). But, all told, Southerners spend less time with radio and television—no doubt to the despair of the advertisers who subscribe to the *Target Group Index*.

These regional differences are not enormous. Region makes less difference than education does, for instance. But regional differences are about the same size as differences between blacks and whites—about the same size, that is, as some other "cultural" or "ethnic" differences in the United States.

Wilbur Zelinsky, a cultural geographer, came to similar conclusions in a 1974 study of state-to-state variation in magazine circulation and membership in hobby and special interest groups. Zelinsky did not assume in advance that the Southern states resemble one another or differ from the rest of the country, but his data revealed that they do, displaying a "Southern" factor independent of a separate urban factor and also (Zelinsky argued) independent of standard of living. Some of the regional differences are fascinating in detail (the South has relatively more readers of *True Confessions* and fewer of *Cosmopolitan*, more readers of *Guns and Ammo* and fewer of *Psychology Today*), but the point that concerns us here is that, on the average, Southerners belonged to fewer of these organizations and read fewer of these magazines.

Similarly, it is well known among political scientists that Southerners have lower levels of political activity—notably voting, but activities of other sorts as well. The literature on this subject tends to explain these lower levels with reference to such political institutions as disfranchisement and one-party politics, and surely these factors do explain many of the South's behavioral peculiarities. But even if these institutions were absent, Southerners might still display relatively low levels of political activity. That would be consistent with their behavior in other spheres of life, and the explanation would have to be sought outside politics.

Let me mention, finally, one other sort of activity. A while back, the National Endowment for the Arts commissioned Peter Marsden and me to do a study of Southerners' behavior with regard to the arts, both as participants and as "consumers." The Endowment was concerned, I believe, that so little of its money was being spent in the South (the president at the time was from Georgia), and it wanted to know if there were "barriers" of some sort that kept Southerners from participating in the arts as much as they would like. We examined a number of surveys on the subject and concluded that barriers are no more of a problem in the South than anywhere else. Southerners simply don't want to do as much of what the Endowment likes to pay for as New Yorkers or Californians do.

Once again, we found Southerners doing less of most things. Out of forty activities that the Louis Harris organization asked about in 1973 (about two-thirds of them having to do with "the arts," loosely defined), there were only three that Southerners were more likely than non-Southerners to do: sing in a choir, listen to religious music, and listen to country and western music. If the Endowment really wants to spend more money in the South, apparently it should subsidize Baptist churches and honky-tonks. The South is no longer the "Sahara of the Bozart," but it very much looks as if most Southerners are getting as much high culture as they want.

Perhaps it is time to generalize. Although there are individual Southerners who are hyperactive, by any standard, and there are activities (going to church, for instance) that the average Southerner does more than other Americans, it seems that the average Southerner is less likely to participate in the average activity.

If we can entertain that possibility, for a while at least, it is worth asking why that is so, if only because there has been no shortage of answers. Most of them are almost certainly wrong—or at least inadequate. H. L. Mencken's view, echoed more recently by the Kentucky-born gonzo journalist Dr. Hunter S. Thompson, was that white Southerners are genetically disposed to idleness and vicious habits. On the other hand, the South Carolina writer Josephine Pinckney argued that these are innate *black* traits, somehow spread to whites by contagion—a proposition not calculated to please much of anybody. Southerners' favorite explanation, or excuse, has probably been the weather. In its old version, favored by Robert Beverley and William Byrd, this argument has it that it is too easy to get by where food grows almost by itself and nobody needs many clothes. A version more applicable to the urban, industrial South says that it is just *too hot* to do much of anything. Either way, there have been many observers ready to echo the railroad tycoon James J. Hill, who said once that no man on whom the snow doesn't fall can be worth a damn. (Mr. Hill, I should note, was born in Canada.) Still another popular theory pointed to the effects—points to the legacy—of slavery, producing laziness in slaves and slaveholders alike (for obvious reasons), and also leading nonslaveholders to believe that exertion was for slaves, and beneath their dignity. Thomas Jefferson and William Byrd both advanced early versions of this argument.

I believe I could show that each of these explanations comes up short, and maybe one day I will write a book on the subject. But I don't know that I or anyone else will ever be able to say for certain where these Southern characteristics *do* come from. Maybe I can say something, however, about why they have persisted.

Whatever their origins, Southern attitudes toward work and leisure have been picked up and made part of our regional subculture. Like other cultural traits, these are not things that individuals work out for themselves. They are learned from those around us while we are growing up and from each other after we are grown. A large part of any culture is made up of shared views of what is appropriate; success in any culture depends on learning what those views are and, ordinarily, on coming to share them. We are not born knowing what to do with our leisure time, nor does each of us make it up as he goes along. Like calling your elders "sir" and "ma'am" instead of "buddy" and "lady," like putting butter and salt on grits instead of cream and sugar, we learn how to pass time appropriately. People who share the same culture—a regional culture, for instance—learn more or less similar lessons.

Recall that the regional differences we have discussed are relatively small. Although they are consistent and apparently reliable, they are of about the same order of magnitude, as I said earlier, as are other "ethnic" differences in the United States. We can presume that they are passed on from generation to generation like those other differences; certainly, like them, they have proved to be surprisingly adaptable and persistent. These differences were still evident after it stopped being easy to make a living in the South (if it ever was); and they are still here now that it is becoming easier again. These differences survived the abolition of slavery, and they have survived the eradication of malaria and hookworm infestation. They have persisted through long-term changes in the Southern climate, and through the introduction of air-conditioning. It will take more than new technological gimmicks, new kinds of jobs, or moves from the country to the city to eradicate them.

To be sure, these changes may allow old tastes to be expressed in new ways. Louis Rubin has written a witty discussion, for instance, of how CB radio has done away with the loneliness of the long-distance driver, providing a new medium for the sort of aimless conversation that used to be conducted on front porches. (And anyone who doubts the regional provenance of CB should listen sometime to its lingo spoken with a New York accent.) Even if we were to see changes in behavior, in other words, it would not follow that values had changed. Indeed, in a changing environment the same values *should* produce different behavior.

Let me conclude by examining one final set of survey data. Someone might reasonably ask what it is that Southerners do more of. After all, the Southern day has twenty-four hours, just like everyone else's, and Southerners must be doing something when they are not consuming, driving, watching television, reading magazines, participating in clubs and hobbies, or engaging in political or cultural or artistic activity. The data we have looked at so far put a new twist to the famous inquiry from *Absalom, Absalom!:* "Tell about the South," Shreve says. "What's it like there? What do they *do* there?"

The NEA study I mentioned earlier set out specifically to find things that Southerners do more of than other Americans, and in a Harris poll from 1978 it did locate a number of them. Here are a few:

fixing things around the house
helping others
eating
developing one's personality
having a good time with people close to you
resting after work
getting away from problems
taking naps
and
"just doing nothing"

It seems to me that most of these are not inherently unworthy ways to pass the time, however difficult it may be to market them. In another context, one historian has suggested a way to summarize them. The South, Ben Hunnicutt suspects, has a pattern of leisure that is more "time-intensive," less "goods-intensive." Southerners have been likely to choose activities that take more time, but less money, than those chosen by other late-twentieth-century Americans. If Hunnicutt is right, here is another way the modern South is hanging on to elements of a preindustrial, "folk" culture—as the historian David Potter said it has.

Another way to describe it is the way John Crowe Ransom did sixty years ago in *I'll Take My Stand,* the manifesto of the Vanderbilt Agrarians who called themselves the Twelve Southerners. Responding implicitly to the charge that the South had neglected the arts (as the National Endowment apparently believes it still does), Ransom wrote that the arts of the South are the "social arts of dress, conversation, manners, the table, the hunt, politics, oratory, the pulpit[;] arts of living and not arts of escape; . . . community arts in which every class of society could participate after its kind." "The South," he concluded, "took life easy, which is itself a tolerably comprehensive art."

My Tears Spoiled My Aim

Violence in Country Music

■ My tears spoiled my aim; that's why you're not dead.
I blew a hole in the wall two feet above the bed.
I couldn't see where you were at, my tears were fallin' so.
I tried to shoot by ear, but y'all were lyin' low.
(© 1982, Bozart Music)

Thus, the chorus of one of the great unrecorded country songs of our time.

As a matter of fact, I wrote it. Perhaps it says something about country music that when I sent this song to a friend with music-industry connections, he wrote back that it was "too cerebral." I had hoped it was at least gastrointestinal, if not cardiovascular, and I still think it is a natural-born hit. Any country superstar who wants to record it should drop me a line. We can work something out.

Country music has a violent strain, and it comes by it legitimately. As Bill Malone has pointed out, Ralph Peer's famous 1927 recording session in Bristol brought together two musical streams that eventually gave rise to country music as we know it. From the hill country of Southwest Virginia came the Carter Family, working in what can be recognized as an extension of the old English and Lowlands ballad tradition preserved in the Southern mountains for a century and a half or more. And out of the Deep South, from Mississippi, came Jimmie Rodgers, singing music so heavily influenced by traditional black forms that some of his listeners assumed that the Blue Yodeler was himself black. (A few decades later, his fellow Mississippian Charlie Pride was subject to an amusingly similar misapprehension.)

Given this genealogy, perhaps we can take it as axiomatic that anything the white music of the Southern uplands and the black music of the Deep South have in common will be found double-distilled in the hybrid that is modern country music. Certainly that is true of the sort of theme that my song exemplifies, in its cerebral way.

Both of country music's lines of descent treat violence as a normal, understandable, almost commonplace feature of adult life. While white Southerners like Tom Dula and Poor Ellen Smith dutifully carried on the blood-soaked tradition of Mary Hamilton, The Two Sisters, Lady Margaret, Barbara Allen, Lord Thomas and Fair Ellender, black Southerners like Frankie and Johnny, Stagolee, and Railroad Bill embodied a tradition of separate-but-at-least-equal violence.

In this respect, country music and its ancestors are strikingly different from American popular music in general—or perhaps we should say that American popular music differs from them, and from the traditional music of much of the world. In any case, you can listen to American popular music for days without encountering so much as a fistfight, much less a killing. Top-40 America is a very different place from the world of "The Banks of the Ohio," the world of "That Bully of This Town," or the world of modern country music.

In part, at least, this is because the world of New York and Los Angeles is still different from that of Nashville, Austin, and Bakersfield. Country music retains an identification with the white working-class South, and it still largely reflects the values and assumptions of its performers and its audience. With a couple of exceptions, violence in country music mirrors violence in what passes for real life in the South. In particular, in music as in real life, Southerners' rates of homicide and assault greatly exceed the national average. And, as in real life, violence in country music is culturally patterned, seen as legitimate (or, anyway, understandable) in some circumstances—but only in some.

The locus classicus of traditional homicide, for instance, is what the FBI calls "romantic triangles and lovers' quarrels." This sort of killing has historically been twice as common in the South as in the rest of the United States. It is of course conspicuous in country music's musical ancestors, and it is common in modern country music as well. When Waylon Jennings sings in "Cedartown, Georgia" about killing his love and sending her body home, the classic theme of love, betrayal, and revenge is not at all disguised by the modern accretions of pawnshops, motels, trains, and cheap pistols.

When songs of this sort are at all novel, it is a novelty of accident, not essence. The basic theme is archetypal, not to say hackneyed—although Kenny Rogers worked a successful variation in "Ruby," a song about a paraplegic Vietnam veteran who tells his two-timing wife that he would kill her, if he could move.

As in the ballad tradition, the narration of these songs is frequently first-person, and the tone somber, not to say grim. One cheerful exception, however, shows that well-internalized social norms can become virtually reflexive: in "Kate," Johnny Cash complains to his dead wife that it is her fault he is in prison, because "your cheatin' pulled the trigger."

When country songs in this tradition are serious, they are generally true to their musical heritage and to the facts of Southern life. But they depart from both in one important way: what the police call "the perpetrator" is rarely a woman. Country Johnnies do their Frankies wrong all the time, but rarely pay for it like Johnny, with their lives. (One exception is the rhythm-and-blues-influenced "Betty's Being Bad," by Marshall Chapman, one of the relatively few female country songwriters, in which Sawyer Brown observes that a ".45's quicker than 409. / Betty cleaned house for the very last time.") Women complain a lot in country music, but for the time being "Stand by Your Man" seems to be the watchword. The Other Woman may be a likelier target—Loretta Lynn's "Fist City" suggests as much—but even this is a fairly new development.

And when we examine violence outside the realm of good love gone bad, we enter an almost exclusively masculine world. Here the targets run an impressive gamut. In fact, the men of country music have beat up at one time or another on most things that move and some that do not. In another Johnny Cash song, for instance, the singer threatens to stomp the head of an "Egg-suckin' Dog" in his henhouse; in a Ray Stevens talking-blues number the victim is a vending machine that steals quarters, shot with a .45 in the "Thank you" sign; and, naturally, "Bubba Shot the Juke Box." But of course the most frequent targets of country-music violence are other men, and a readiness to fight when necessary is an important part of the definition of country manhood. Hank Williams, Jr., suggests as much almost incidentally in "Texas Women" (he doesn't ride bulls, but he has fought some men), and Kenny Rogers puts it more explicitly in "Coward of the County": "Sometimes you have to fight when you're a man." In country music, a man has to fight when his honor is at stake, and he is allowed to fight (a rather different proposition) when someone threatens his pride, person, or property—or at least seriously annoys him somehow.

That may sound as if it includes everything, but it does not. Like Southern culture in general, country music does not condone violence in which the victim was in no way "asking for it." That kind of violence calls not for understanding but for retribution (often violent itself, of course). This is evident even with dogs and vending machines. A man who kills the one for pleasure or vandalizes the other for profit should not look to the country-music audience for sympathy, but if they suck his eggs or steal his quarters, it is a different matter.

"Coward of the County" offers another case in point. The gang-rape of the hero's sweetheart is bad violence—so much so that the hero must break his oath not to fight. Just so, in "A Country Boy Can Survive," by Hank Williams, Jr., the singer's friend was knifed to death on a city street for $43.00. This sort of homicide, still relatively rare in the South, is viewed as outrageous. It cries out for vengeance—that is, for good violence—and the singer would like to oblige, he says: he'd like to spit tobacco in the mugger's eye and shoot him dead.

Occasions for violence certainly include the chivalric defense of womanhood—avenging gang-rape, for instance, or responding to a daughter in trouble, as in Rodney Crowell's "Leaving Louisiana in the Broad Daylight":

> Did you ever see a Cajun when he really got mad,
> Really got trouble, like a daughter gone bad?
> It gets real hot down in Louisiana.
>
> A stranger better move it or he's gonna get killed.
> He's gonna have to get it or his shotgun will.
> It ain't no time for lengthy speeches.

But there is considerable overlap between the defense of women and the defense of one's self-respect, or even of one's chattel. Sometimes it is hard to tell where one leaves off and the other begins. And sometimes it is not hard at all, as when someone tampering "with the property of the U.S. Male" is warned that his head may wind up "about the shape of a stamp." (In this song, recorded by both Elvis Presley and Jerry Reed, the singer's woman is apparently passive, so she is not a target herself.)

The linkage between machismo and patriotism is only implicit in "U.S. Male," but a few other songs have brought it front and center. Insults to one's country, like insults to one's woman, are taken personally, as in "Walking on the Fighting Side of Me," Merle Haggard's warning to Vietnam War protesters, or "In America," in which Charlie Daniels warns "all you outside people" to leave America alone, or else. ("In America" represents a considerable ideological migration for Daniels from his early "Uneasy Rider." That was also a violent song, but it implied that long hair and superpatriotism could not go together, and sided with the longhairs.)

Like country music itself, however, country-music violence is seldom explicitly political. It rarely takes place in the streets and certainly not on the barricades. But it often does take place more or less in public, especially in the kind of dive David Allan Coe describes in "Long-Haired Redneck," the kind of place where bankers gape at cowboys, cowboys make fun of hippies, and hippies pray they'll get out alive. In this setting it is understood that men fight for all sorts of reasons, or for none at all.

In Charlie Daniels's "Trudy," for instance, the violence results from a back-room gambling dispute. The Cajun hero proudly reports that he gave as good as he got: when he accused another gambler of cheating and the man reached for a gun, he hit him with a chair and ran for it; it then took "half the cops in Dallas County" to put him in jail. Another grotesque example from among scores, if not hundreds, is sung—or rather recited—by Johnny Paycheck. When a bully ill-advisedly spits "Colorado Kool-Aid" (Coors beer) in the ear of the singer's Mexican friend, the Mexican pulls a knife, cuts off his assailant's ear, hands it to him politely, and suggests that next time he wants to spit beer he should spit it in his own ear. As in these last two examples, incidentally, Mexicans and Cajuns figure prominently among "perpetrators" in country music. I have no idea what that means, but if these are ethnic stereotypes, they are at least grudgingly admiring ones.

In fact, when a character in country music is as tough as he can legitimately be, when it is plain that he is not going to take any aggravation from anybody, we find ourselves dealing with something of a heroic figure. Unlike the bad man of the white ballad tradition, this redneck hero does not urge us from the gallows to learn from his example. He is an unrepentant outlaw, a countercultural figure, a whiteface version of the "bad nigger" of traditional black music: Railroad Bill, Stagolee, Eddy Jones, Bad-Lan' Stone—that Bully of This Town, by whatever name.

Although songs of the redneck hero seem usually to be set in a respectful third-person, occasionally he speaks for himself. When someone makes remarks about David Allan Coe's earring and long hair, for instance, Coe sings that somebody ought to warn the man that he, Coe, has been in prison and will punch him out if he keeps it up. In what some of us take as a metaphor for the South as a whole, Coe sings that his long hair does not cover up his red neck, and (the metaphor breaks down) that he has won every fight he has ever fought.

Like the songs of the black tradition they mimic, these songs are often humorous, and a few are downright Paul Bunyanesque. Consider, for example, Jerry Reed's "Amos Moses" (another Cajun, by the way), who hunts alligators with one hand—all he has left, thanks to his dangerous, and illegal, occupation. In "Amos Moses" a sheriff goes into the swamp after the hero but never comes out, a disappearing lawman reminiscent of the moonshine-seeking Treasury agents in "Rocky Top."

For really reliable humor, though, we must turn to another stock character often found fighting his way through country music—the amiable, joshing, self-deprecatory figure of the good old boy. In "Just Good Old Boys" Moe Bandy and Joe Stampley present the classic description: men who fight with their bosses and lose their jobs, get thrown out of bars, steal city trucks and wreck them in the mayor's yard, beat up their brothers-in-law over football bets, get shot at for fooling around with married women, but, other than that, are "just good old boys."

These fellows may be fools, but they are obviously not cowards; they will fight if they have reason to, and (with willing antagonists and alcoholic encouragement) sometimes they fight just for the hell of it, to see who wins. Unlike the invincible redneck hero, though, sometimes they lose, as in "Holdin' the Bag," in which Moe and Joe wax nostalgic about an occasion when Joe told another man that Moe wanted to fight—and almost got him killed. The singer is also the loser in "Who Was That Man That Beat Me So?" by Johnny Paycheck, in Ray Stevens's "Blue Cyclone" (the nom de guerre of an infuriated professional wrestler), in the last verse of Hank Williams, Jr.'s "Attitude Adjustment" (courtesy of the police, with nightsticks), and in a great many other recent examples.

In all these songs the humor relies on the victims' having brought their troubles on themselves, by ill-considered actions or remarks. Unprovoked violence wouldn't be funny; in these cases, however, the singer-losers may be rueful but are really not entitled to complain. Thus, even songs by losers about losing seldom question the traditional understanding that violence can be an amusing spectacle, a wholesome form of recreation, and an appropriate way to redress grievances. Indeed, they rely upon it. To understand the functions of violence in country music, one could do worse than to read Bertram Wyatt-Brown's *Southern Honor.* Wyatt-Brown argues that the acids of modernism have undone that cultural pattern; if so, however, they have barely begun to work in country music.

But a few recent songs do reveal some corrosion. Bobby Bare's "If That Ain't Love," for instance, gives us a reductio ad absurdum of the romantic triangle song:

> In front of your apartment I waited all night long
> To see who you come home with and if he done you wrong,
> Then I slapped your face and I stepped on his—
> If that ain't love, what is? ©

Similarly, when Bare sings about "Big Dupree" (yet another Cajun), he gives us a sort of down-home Heloise and Abelard story that pretty well defines the limits of the loser's point of view:

> There ain't nothin' worse 'n
> What he done to my person
> And I spend the evenin's cursin'
> Him in [the next words are sung falsetto]
> slightly higher tones. ©

The lyrics of these two songs, like those of Bare's "Drunk and Crazy" ("Gonna let the good times roll"), were written by Shel Silverstein, who also wrote Johnny Cash's hit "A Boy Named Sue," another classic of comical violence. These songs reflect a thoroughly "modern" mentality, one that finds traditional country-music attitudes toward violence more amusing than violence itself. They are parody, doing with violence what Silverstein's "Put Another Log on the Fire" (sung by Tompall Glaser) did with sexism.

Of course, parody is a difficult form. Some country songs written in dead earnest are so bizarre that parodies do not have to be understood as such. If they are not so understood, it may be better for sales, but any critical function of the songs is lost. Let me close, though, by mentioning three other songs, where the criticism of traditional patterns is less oblique and the point harder (although probably not impossible) to miss. All deal with the sort of barroom brawling that figures so prominently in country music, the redneck's specialty and the good old boy's Saturday-night avocation.

In "The Winner," a young hotshot challenges an old and undefeated brawler. The veteran shows all of his scars and describes his various operations in detail, says that he is tired of being a winner and that the younger man can be one now, and turns away. In "Goin' Back to Texas," on the other hand, the aging narrator acknowledges that he is over the hill, but avers that

> I can dance and I can fight and throw a whiskey glass—
> I'm goin' back to Texas and be one more horse's ass. ©

Finally, the "Redneck Mother" is ironically applauded for having raised a son who, at thirty-four, still drinks in taverns, "just kickin' hippies' asses and raisin' hell." In each of these songs, it is *age* that seems finally to have made violence stupid, contemptible, or pathetic.

All three songs, as it happens, have been recorded by Bobby Bare, who almost seems to be specializing in this sort of thing. It is interesting that recording such material does not seem to threaten his cultural bona fides: nobody, so far as I know, has called him a wimp. But none of these songs was a major, top-ten hit, and half a dozen songs by one performer hardly constitute a trend. Whether Bare's skeptical view reflects the "acids of modernity" at work or just the mellow wisdom of his own middle age, such skepticism remains a scarce commodity in country music.

Playboy's Southern Exposure

With Athan Manuel and Charles R. Wilson

■ *Playboy* magazine has seldom been taken seriously as a cultural force in American life. Its founder, Hugh Hefner, has of course taken it very seriously indeed, and there have been other exceptions: Gay Talese in *Thy Neighbor's Wife,* for instance, or more recently the feminist critic Barbara Ehrenreich in *The Hearts of Men.* These writers attribute to *Playboy* an enormous and fateful impact on American sexual mores, for better (Hefner and Talese) or for worse (Ehrenreich), but it still seems an open question whether the magazine actually sparked those changes or merely profited from them.

Nevertheless, there is little doubt that *Playboy* provides fodder for anyone who wants to understand the attitudes of Americans—or at least those of many young American men. For that purpose, it is not terribly important whether *Playboy*'s content shapes its readers' attitudes or just reflects them. It matters only that the magazine not be grossly out of step with what its readers are thinking, and its commercial success attests to that. Since a great and conspicuous part of the magazine is devoted to portrayals of young women (a subject of enduring interest to young men), the images of these women that *Playboy* presents are of particular importance.

Shortly after *Playboy*'s thirtieth anniversary, we examined its portrayals of young *Southern* women with a number of questions in mind: Are Southerners overrepresented (as they have been, for example, among Miss Americas and beauty queens of various sorts)? Does the magazine present a standardized, stereotypical image of "the Southern woman," or are there several competing images, or are the portraits so dissimilar that no generalization is possible? Did the presentation of Southern women in *Playboy* change during the magazine's first thirty years and, if so, what accounted for the changes?

It simply will not do to treat this subject too solemnly. In particular, the sort of methodological throat-clearing that usually accompanies a content analysis of popular culture materials seems uncalled-for. To put it as simply and briefly as possible: We examined the first thirty years' worth of *Playboy,* looking primarily at the 350-plus "Playmates of the Month." (Mr. Manuel, the junior author in point of age, did the looking.) The conventions of the Playmate genre require that the magazine say something, but not much, about the subject of the photographs. A few paragraphs purportedly describe the background, life, and views of the model, and provide the student of these matters with small, manageable samples of text, often banal, but suitable for analysis. (Note that it does not matter for our purposes whether the text presents a representative selection of genuine facts about the model or a completely fictitious persona. To avoid an excess of skeptical adverbs, we shall treat these sketches as if they were all, in fact, factual.) We also examined a 1977 feature entitled "The Girls of the New South," which gave *Playboy* an opportunity for a more expansive treatment of Southern women.

Since the Playmate format allows for only the most basic information about each model plus a very few individuating details, it is significant that nearly all the texts we examined indicate where the models grew up or currently live. Evidently these geographical data —like age, occupation, and (in some circles) astrological sign—convey a great deal of implied "information" and are particularly important in situations like this one, where communication is artificially constrained.

Fewer Than Expected

From a regionalist's point of view, the most striking datum is the relative absence of Southerners among *Playboy*'s Playmates. Far from being overrepresented, Southerners have been dramatically *under*-represented. As the accompanying figure shows, during the magazine's first twenty years, among its first 250-plus Playmates, only six

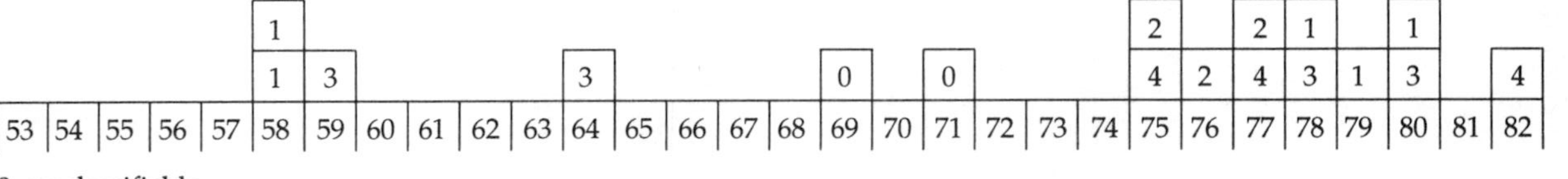

0: unclassifiable
1: bathing beauty
2: starlet
3: belle
4: good old girl

Frequency and Type of Southern Playmates, by Year

had any connection with the South at all. Southerners were more common in *Playboy*'s third decade, but even during 1975–1980, the peak period for Southern Playmates, fewer than 14 percent of all Playmates were from the South, far fewer than the 30 percent or so that one would expect if region had nothing to do with the matter. In part, this surely reflects the exigencies of recruitment: in *Playboy*'s early years many Playmates were from Chicago, where the magazine's headquarters were located; now the most frequent area of origin seems to be California. But might this also reflect persisting regional differences in mores? Probably Miss America is (or was until recently) more honored than the Playmate-of-the-Year in every American region, but perhaps Southerners have tended to see the difference as greater than other Americans have. If so, that would help to explain why young women from the South have been as underrepresented in the one category as they have been overrepresented in the other.

Whatever the effect of regional culture, however, the chart suggests another factor at work: the image of the South in the minds of non-Southern Americans. Between 1965 and the early 1970s, at a time when the South's press was even worse than usual and its associations for most non-Southerners were probably unpleasant, only two Playmates were from the South. Both were Texans, from the geographical fringe of the region; in neither case were the model's origins emphasized; and one of the two (Miss October 1969) was also *Playboy*'s first black Playmate. A few years later, however, came a brief interlude that George Tindall has characterized as a period of "Southern-fried chic." Sam Ervin had appeared on television screens and T-shirts; Burt Reynolds was dominating the screens of drive-ins around the country; the electorate was feeling the first pangs of what John Egerton calls "peanuts envy" and preparing to elect Jimmy Carter to the White House. At exactly that time *Playboy*—ever with-it—began to present young Southern women as Playmates of the Month. Within three months of Carter's inauguration, *Playboy* was on the stands with its "Girls of the New South" feature.

Immediately after Carter's fall, on the other hand, Southern Playmates quickly went the way of Senator Sam T-shirts. There were none at all in 1981, and only one, a Texan, in 1982. Plainly there are fashions in pinups, as in automobiles and clothing, and *Playboy* has been on top of the situation right along.

Four Types

This brief review, however, conceals as much as it reveals about *Playboy*'s portrayal of Southern women. There have been several different *types* of Southern Playmates—familiar, stereotypical categories that subsume nearly all of the young Southern women who appeared as Playmates during the magazine's first three decades. We can identify four, which we shall call the *bathing beauty,* the *starlet,* the *belle,* and the *good old girl.*

Only two Southern Playmates from this period do not fit comfortably into one or another of these four categories, and it is interesting that they are the two Texans mentioned earlier who appeared in the magazine at the nadir of the South's recent reputation. Jean Bell, the black Playmate from 1969, was virtually sui generis. Although she came from Houston, she was not presented as particularly Southern, particularly Texan, or even particularly black. The headline of the feature was "Black Is Beautiful," but the text emphasized that Ms. Bell was not involved in politics and stressed her love for bowling. The other Southern Playmate from the decade 1965–1974 was Karen Christy, Miss December 1971, from Abilene, Texas. The text mentioned only that she had been a commercial art major at North Texas State before dropping out to become a secretary in Dallas. Texas and, by implication, the South were not presented as places of any unusual interest.

Playboy was attempting at the time to be youth- and student-oriented (with frequent derogatory references to people over thirty, for instance), and we can surmise that its editors feared that their desired audience would not believe a romanticized version of the South and not enjoy a gritty, "realistic" treatment. The upshot was that they simply presented no image of the South at all. With these two exceptions, however, all the Southern Playmates we examined fit into one of the four categories we have identified.

The Bathing Beauty

But about half of them fit one of the two types that are not distinctively *Southern.* The bathing beauty, for example, is an old type (she seems to have appeared about the same time as photography) and in no way uniquely or even predominantly associated with the South. The first two Southern Playmates (two of the three from the magazine's first decade) both belonged to this well-established category. Miss August and Miss December 1958 were both from Miami, and both loved sun and surf. Virtually the same text could have been used—and was—for many West Coast Playmates. When Southern Playmates reappeared in the middle and late 1970s, three can be identified as simply throwbacks to this type. All three, however, were from Texas (although one had moved to Florida), indicating that the "Redneck Riviera" of the Western Gulf Coast had begun to flourish, and that the Sunbelt no longer consisted only of Florida and California.

There is some question whether any of these bathing beauties, early or late, should properly be regarded as Southerners at all. Not only do all come from the edges of the region, Texas and Florida, but two of the three from the 1970s were described as "army brats"—fauna of the modern South, to be sure, but birds of passage. One of the army brats did list *Gone with the Wind* as her favorite book, which with her given name (Louann) almost makes up for her strong environmentalist views and opposition to litterbugs. But the other army brat, Gig Gangel, could just as well have been a Californian: she is supposed to have liked blue eyes, surfers, Ferraris, and sunsets—and *her* favorite book was *Looking Out for Number One.*

The Starlet

Another type of Playmate Southern only by location is represented by three specimens from the period 1975–1977. These women were all Southern expatriates to California, and all three were what used to be known as starlets. Their Southernness was almost incidental, and the text accompanying their photographs was very similar to that for other starlets from, say, the Midwest.

Star Stowe, from Little Rock, had so far turned her back on her Southern (hence probably Evangelical Protestant) roots that she had become a witch. Laura Misch of Tulsa, however, seemed to be having trouble adjusting to the karmic scene: when asked her sign,

Miss February 1975 replied "Exxon." Ms. Misch, whose road to Hollywood took her via that latter-day House of the Rising Sun, the New Orleans Playboy Club, did use her regional background to good advantage in her first movie role, as a prostitute in *Mandingo.* (One picture showed her being dressed by a black extra playing a maid.)

Two cases hardly establish a trend, but the third Southern starlet, Daina House from Dallas, also made her film debut as a prostitute (in *The Winds of Autumn*). She allegedly wrote the text that accompanied her *Playboy* pictures herself, but she did little credit to the Southern literary tradition, and most of what she had to say is too boring to repeat. She employed the ultimate Southern metaphor, however, when she asserted that she was "ballsy, but I'm made of cotton inside and hurt easily."

The Belle

Rather more interesting than the bathing beauty and the starlet is the third of our four types, the Southern belle. This type is well known in American popular culture outside the pages of men's magazines and might seem at first an easy one to adapt to the Playmate format. But there are some problems with it, and some of them are illustrated by the first Southern Playmate to break out of the bathing beauty category, Nancy Crawford, Miss April 1959.

As a Virginian, Miss Crawford was, inevitably, photographed at a fox hunt, clad in impeccable cap and pink. She looked, alas, implausible. Whatever else she may be, the belle is undeniably a lady—and ladies do not take off their clothes for photographers from Chicago. Moreover, this rendering of Virginia aristocracy was hardly in keeping with *Playboy*'s "girl-next-door" policy: the kind of man who reads *Playboy* is unlikely to move in those circles.

After a five-year hiatus, *Playboy* tried again, with a somewhat more middle-class version of the belle. Nancy Jo Hooper from Savannah was not a full-blown bathing beauty, but she did have a mild water fetish: she liked swimming, boating, and waterskiing. According to the write-up, this "Georgia peach" from "the heart of the old Confederacy" also liked the classic Southern activity, or inactivity, of fishing. Otherwise, she was presented as a lady to the core. She liked elegant shrimp dinners, had a "caramel drawl," and was photographed at a historic site (an old fort in the harbor). Her trip to Chicago for the studio photographs was her first trip in an airplane, but for some inexplicable reason her goal in life was to be a nurse in California. Perhaps she felt it was time to leave Savannah.

Ms. Hooper's unaristocratic taste for fishing was a sign of things to come, although they took another decade to arrive, in the form of the good old girl. Meanwhile, *Playboy* made a few more attempts to come up with an unladylike belle. That effort was plainly evident in its 1977 feature on "The Girls of the New South." The article was explicitly tied to the Carter presidency, which had just begun, and *Playboy* almost seemed to be offering a new-style belle—urban, economically independent, and sexually liberated—to go with this new-style Southern politician.

But it built on a base of tradition. *Playboy* even presented the Southern woman's sensuality as a regional tradition, falling back on an old theory to explain it: "Maybe it's those hot, humid nights." The beauty of the women was also treated as a tradition of the Southland—its oldest, the article noted, since the passing of the boll weevil and pellagra—and Scarlett O'Hara made her customary appearance as a model for young, high-spirited Southern women.

Sometimes the article showed the strain of trying to deal with new observations in old categories, but it usually managed. The women in the feature clearly came from an urban world, the Sunbelt South, but they were portrayed in rural and plantation settings. One of them, a migrant from New Jersey, even insisted that Southern discos offered "a plantation atmosphere with lots of wood," while Northern discos were all bright lights and shiny glass.

The traditional Southern attribute most prominently displayed in the article, however, was regional pride. *Playboy* observed that Southerners were "feeling a mite chauvinistic about their magnolia-and-sunshine existence" in those first months of the Carter era, and that, the magazine clearly implied, was an OK way to feel. Throughout the article, the models were quoted in praise of the South, its traditions, its gentlemanly males, and its easygoing life-style. Even the New Jersey migrant pledged fealty to her new homeland, proclaimed its superiority, and traded in the imagery of an earlier South.

All these contradictions and tentative conventions were plainly evident a little over a year later, in June 1978, when the Playmate of the Month was another "new fashioned Southern belle." Gail Stanton smelled like jasmine, came from Memphis, and echoed a number of themes then being heard in White House rhetoric and the lyrics of country songs like Tanya Tucker's "I Believe the South Is Going to Rise Again." She loved the South and believed that young Southerners were going to make it a better place, solving the problems of the past. Ms. Stanton's tastes were a mixture of the traditional and the modern: her favorite movie was *Gone with the Wind;* her favorite book, *Vivien Leigh: A Biography.* But her regional pride sometimes showed a xenophobic edge that was as unladylike as it was odd in a "new-fashioned" Southerner of the Carter era. She said Northerners were the most unfriendly people she had ever met, and she hated Northern food: "Northern food makes me sick. I wish they knew how to cook blackeyed peas and rice and red-eye gravy and grits." Ms. Stanton had once dated Elvis, which put her Southern bona fides beyond question, but also further indicated that she was probably not a lady.

This new image of the Southern belle, like the then-current new image of the South, was unstable, short-lived, and perhaps only a media creation in the first place. In March 1980, *Playboy* reverted to tradition—a sign, for those with eyes to see, that Jimmy Carter's days were numbered. Henriette Allais was a "Southern Comforter" photographed in a plantation setting and wearing a hoop skirt. "You might call me the quintessential Southern belle," she suggested, lest *Playboy* readers miss the point. Her favorite drink was lemonade, and her favorite movie goes without saying. "Scarlett and I are very similar," said Ms. Allais, an orthodontist's assistant. But the creeping Californication of the South was evident in her list of Good Things: yoga, exercise, the smell of spring, and the face of a happy child.

The Good Old Girl

Scarlett O'Hara, some may feel, has a lot to answer for, but she also found the role of Southern lady somewhat restrictive and laid the groundwork for another version of Southern femininity, one that seems increasingly popular with *Playboy:* that is, the good old girl, often from Texas.

The first Playmate of this type anticipated the Carter presidency by more than a year. Miss April 1975, Victoria Cunningham, was originally from Beaumont, had been kicked out of stewardess school in Dallas for being too rowdy, and moved to Los Angeles, where she apparently hung around the Beverly Hills Tennis Club until a *Playboy* photographer showed up. Ms. Cunningham would have been a natural in any of a half-dozen Burt Reynolds and Jerry Reed films, and one wonders why she never appeared in any. Perhaps she did. In any case, like Scarlett O'Hara, she enjoyed pistol shooting, with a special fondness for .45s and .357 magnums. She said it was "fun to hit the target."

Two years later, in 1977, *Playboy* found another Beaumont sharpshooter, Debra Jo Fondren, although her tastes ran more to skeet shooting than pistolry. Ms. Fondren lived in a log cabin, and her biographical notes contained what cognoscenti will recognize as nostalgic references to the traditions of twenty years earlier: she liked swimming, sunbathing, and waterskiing, on the one hand, and horseback riding on the other (although, as a Texan, she preferred bareback to English-saddle, fox-hunt-style riding). Her ambition, evidently realized, was to become a model, although presumably she intended to stay in the South, since she disliked pushy, loud, opinionated people, and being rushed. Perhaps it should have gone without saying, although it did not, that she was not a feminist.

Probably the apotheosis of the good old girl was Miss January of 1982, Kim McArthur, a "Southern Star" from Dallas. Ms. McArthur was a living exemplar of the "Texas chic" evident in the nation's fascination with J. R. Ewing, Willie Nelson, and urban cowboyism—although *Playboy* uncharacteristically got there a year or so after the trend had peaked. Ms. McArthur's principal achievement before appearing in *Playboy* was to be one of the last cut from the Dallas Cowboy cheerleaders, which might have had something to do with her habit of buying fried chicken and eating it under the stars in the middle of empty football fields. Self-described as playful and a romantic, this good old girl was also a chili fiend and found the Lone Star Chili Parlor comparable to anything in Paris or Venice. She did not indicate a favorite book.

The Shape of Things to Come

With the good old girl, *Playboy* may at last have found a formula for selling the Southern Playmate, one with more vitality and more current market appeal than the belle, whose possibilities have been pretty thoroughly played out and whose plausibility in the *Playboy* context always left something to be desired. Whatever her future in real life, in the pages of *Playboy* the belle seems to have gone the way of the Beatnik Playmate, gone—one could say—with the wind.

The good old girl, on the other hand, seems to be alive and well—and not just in the centerfold of *Playboy* but in the lyrics of country music songs, on television programs like "Hee Haw" and "The Dukes of Hazzard," and in innumerable movies that were never nominated for Oscars, as well as one—*Thelma and Louise*—that was. And she is something new. Her male counterpart, the good old boy, has been around, better established, for a longer time: he is pretty much the same type that W. J. Cash wrote about in *The Mind of the South* as the "hell-of-a-fellow." But although the good old girl was dimly foreshadowed in the popular culture of the interwar period and was somewhat more evident during and immediately after World War II, she did not really emerge as a recognizable type until the mass media identified and validated her in the 1970s.

This is not to say that the good old girl has not always been around in real life, in some form or another, but like the original good old boy she was primarily a working-class phenomenon. What the media have done is to make the good old boy and girl what the gentleman and the belle already were: "social types," as the sociologist Orrin Klapp has called them—half stereotypes, half well-defined roles that individuals can choose to play. Increasingly, the good old girl is a role available even to those young Southern women who do not come by it naturally, when they do not feel like being belles.

Of course, they can reject both types and choose some other pattern of behavior altogether, but they should not underestimate the power of the media to spot trends and to identify—perhaps to trivialize—emerging social types. This may already be happening with *Playboy*'s treatment of Southern women. After the cutoff date for our study, one of the Playmates of the Month was a Floridian who did not fit any of the types we identified. As a stockbroker, she could not possibly be a belle or a good old girl; she was obviously not a starlet, and she lacked the bathing beauty's single-mindedness about sun, sand, and surf. In her attributes, her attitudes, and her "dress-for-success" clothes, she was clearly a young, urban, professional woman—in other words, a Southern yuppie.

Thoughts on the Southern Diaspora

■ Sometime in the last few decades the South finally "rejoined the Union." Everyone seems to agree about that, amateur South-watchers and professional Dixiologists alike. If that is so (and I think it is), it is fun to speculate about when the critical moment came. When did the accumulating quantitative changes finally add up to irreversible, qualitative transformation? Any specialist will have his own candidate for when that happened.

An economist, for instance, might pick the instant when the South's per capita income passed the level that the World Bank uses to mark the difference between "developed" and "less developed" countries. By that standard, the South ceased to be an LDC about the time of World War II.

A fertility demographer, on the other hand, might date the transition a little later. The South's birthrate—which had been high, like those of other poor, agricultural societies—dropped in the mid-1950s to a level below the national average (where it has stayed since).

It was about then (not by coincidence) that the South ceased to be a rural society. By the end of the 1950s more than half of all Southerners were living in cities, suburbs, and towns. Some might point to that as the critical transition, whether with satisfaction or with regret.

But a historian or anyone else who shares Ulrich Phillips's view that white supremacy is what the South was all about will emphasize another dramatic halfway point, one that came a few years later still. During the three years 1963 to 1966, support for de jure segregation became a minority view among white Southerners. The percentage of white Southern parents who completely opposed public school desegregation, for example, dropped from 61 to 24.

And any student of Southern politics ought to emphasize a related set of changes that followed hard on the heels of the Voting Rights Act of 1965. Those changes may be best summarized by a little-known story that I heard from William Havard, professor of political science at Vanderbilt. According to Havard, when Jimmy Carter was campaigning on the Mississippi Gulf Coast in 1976, he said that the Voting Rights Act was the best thing that had ever happened to the South. Reporters were surprised when the biracial audience roundly applauded this remark, and one asked Mississippi Senator John Stennis, who was present, if he agreed. Stennis said he did. The same question was then put to Mississippi's other senator, James Eastland, who was also on hand. Eastland paused a moment, then said that, yes, he agreed, too.

Now, whether or not these men actually believed what they were saying, there's no question that the Voting Rights Act had something to do with what they said, and with a good many other wholesome changes in Southern politics as well. In the three years after it became law, for example, black voter registration in Mississippi increased from 6 percent to over 60 percent of those eligible. The consequences of this change are still working themselves out but certainly include the fact that Mississippi now has more black elected officials than any other state in the Union.

All these economic and demographic and political changes are important. All are related to each other and, whether as cause or effect, to still another turning point, one that came sometime between the early 1960s, when for the first time in a century more whites entered the South than left it, and the early 1970s, when the same reversal took place for blacks. That turning point, the outward and visible sign of a regional transformation, is what I want to talk about here.

Now, this is not the place to test hypotheses or to develop a theory or even a sustained line of argument. What I propose to do is to share a few facts that I find impressive, tell a few stories that I hope are interesting, and suggest a few categories for thinking about this topic.

First, a boring but necessary methodological note. We need to keep in mind that a net migration figure has two components: migration in and migration out. If we don't keep them straight, this can produce some confusion. At midcentury, for instance, more than twice as many whites as blacks were leaving the South, but net black out-migration was higher, because whites leaving the South were somewhat offset by whites coming to it (primarily to Florida), while black migration to the South was statistically inconsequential. (In 1960 less than 2 percent of blacks living in the South had been born outside the region.)

People like to refer to the recent migration "turnaround," and I'll do the same because it's convenient shorthand, but that phrase can be misleading. The turnaround has been partly due to increased migration to the South, to be sure. But it has resulted also, and primarily, from a slowing of out-migration. Other than Southern-born "return migrants," there are still relatively few black Americans moving to the South, for instance. What has *really* changed is that fewer Southern blacks are leaving.

Nevertheless, the distinction between in-migration and out-migration provides a convenient way to break up the topic. So here I'll concentrate on migration out of the South, especially in the past. I'll try to indicate the magnitude and nature of the Southern diaspora, say a few things about its effects on the South, and talk about some different ways that Southern migrants have adjusted—or failed to adjust—to the Northern setting.

Migration out of the South is nothing new. Some of the Old South's most vivid stories are about migrants. Think only of the fugitive slaves whose stories were recorded by William Still as they passed through Philadelphia on the Underground Railroad. Think of antebellum white exiles like the Grimke sisters of Charleston or North Carolina's Hinton Helper, men and women who rejected the South's increasingly rigid political orthodoxy and left for reasons of health. Think of the filibusters, hell-for-leather soldiers of fortune like William Walker, who tried unsuccessfully to conquer Nicaragua.

After the Civil War, the flow accelerated. Black Exodusters left to seek opportunity in the Southwest. Unreconstructed whites sought it in Latin America, taking their families and slaves along. Confederate officers who knew no other trade (or just preferred their new one) commanded armies from Mexico to Egypt. Some of my favorite stories involve Southerners who traded on their real or putative gentility to move among the postwar Yankee nouveaux riches. I'll come back to that topic.

This flight from the South, one of the great folk movements of history, continued unabated into the first decades of this century. The Chapel Hill regionalists and the Vanderbilt Agrarians agreed on little else, but both recognized that the South's economic situation in the 1930s was essentially that of a Third World country, something of an internal "colony" of the United States, exchanging raw materials for the manufactured goods produced by the Northeast and industrial Midwest, under unfavorable terms of trade. One of the many resulting parallels with today's Third World can be found in the South's "export" of its surplus population, and we can use the language of today's newspapers to describe it.

The lack of opportunities for professional and entrepreneurial talent in a poor, rural society led to what is nowadays called a "brain drain." More than fifty years ago, Wilson Gee wrote in *Social Forces* of what he called "the drag of talent" out of the South, as ambitious Southerners of both races left to pursue training and opportunity in the North and West. Meanwhile, displaced sharecroppers and the like, both black and white, fled North to compete for unskilled jobs, especially after restrictive immigration laws cut off the flow of peasant immigrants from other poor, rural societies. Some of these went to stay—some had nothing to go back to—but others went essentially as what the Germans call *Gastarbeiter,* "guest workers," building a stake, sending money home (sometimes), intending to return someday.

Perhaps a few statistics are in order here:

In 1950, seven million Americans born in the sixteen states of the "Census South" lived elsewhere. By 1960, nearly ten million did so.

A third of these migrants were black. In 1950, five of every six black Americans had been born in the "Census South," but one of those five had left. By 1960, nearly one in four had left: more than three million Southern-born blacks lived outside the South.

As late as the 1950s, more than half of the black population outside the South was Southern-born; the 1960 census was the first in which that was not true.

The pattern for whites was similar, if less striking. As I said, white Southern migrants outnumbered black ones by roughly two to one, and, in 1960, nearly one in six Southern-born whites had gone to the North or West. In some parts they were a significant minority presence, especially in the West, where Dust Bowl refugees and other Southerners made up 11 and 12 percent of the population of the Mountain and Pacific states, respectively.

Migration from the South followed well-worn pathways. Most Southern migrants to the Northeast came from the Census Bureau's "South Atlantic" region; most of those in the "East North Central" states came from "East South Central" ones; most of those who ended up in the Mountain and Pacific regions began in "West South Central" states. In short, migrants tended to follow the railroads due north from the South Atlantic and East South Central states, and west from the Southwest. (What is due north of Texas and Oklahoma, after all?)

There has been some attention to the effects of the Great Migration on the communities for which the migrants were bound, but little has been written about what this migration meant for the communities they left behind. Natural increase was robust enough to insure that few Southern communities were depopulated by migration, but migration must have had serious demographic, cultural, and economic effects.

Consider who left. Migrants were not a cross section of the Southern population. They were, of course, more likely than average to be black. They were also likely to be young, especially young men. Among whites aged twenty-five to thirty in 1960, net migration out of the South during the previous five years had been 4 percent for males and 1 percent for females, while older whites showed net *in*-migration. For young blacks, net out-migration was 7 percent for males and 6 percent for females, compared to rates less than half as large for older blacks. Migration left behind women, children, and the old. For a heavily affected Southern community, the result was a "population pyramid" like that for a society after a war.

One implication was for political economy. The proportion of the South's population that was economically unproductive and dependent was already high because of the region's high birthrate, but migration made it higher still. With a quarter of the nation's population, and a fifth of its tax revenues, the South was educating a third of the nation's children. Southerners spent a higher proportion of their incomes on education than other Americans, then saw much of that investment flow north when schooling was completed.

Not that the South got nothing in return. Like European guest workers today, Southern workers in the North sometimes sent money back home. Such remittances undoubtedly helped both household and regional economy, although we lack the statistics to let us say how much. But some migrants were very generous: my university's campus, for instance, is littered with monuments to distinguished Southerners who made their names and fortunes elsewhere.

The migration statistics also had implications for family life. "Matriarchal" families, with fathers seldom present or absent altogether, are found throughout the economically underdeveloped world (they are by no means unique to today's black ghettos), and the statistics suggest that such families must have been common in the rural and small-town South at midcentury. True, in strictly economic terms it was better to have the menfolk employed somewhere else than unemployed at home—remember those remittances. But Southerners know that man does not live by bread alone, and this is a topic that awaits the attention of the "new social history."

Migration inevitably affected the South's race relations, too, although it is hard to say whether for better or for worse. It did drain the South disproportionately of poorly educated whites, and replace them to some extent with better-educated in-migrants. Between 1955 and 1960, for example, there was a net in-migration for whites with eight years of education or more, and a net out-migration for those with seven or fewer years of schooling. Since educated whites generally expressed less racial prejudice than poorly educated ones, this may have improved Southern race relations—leave aside what it did for race relations outside the South.

On the other hand, survey research on the subject tends to show that whites who had left the South were less prejudiced than those who had stayed, despite their lower average level of education. It is unclear how much this reflects self-selection and how much it reveals attitude change after moving from the South, but it does suggest that migration may have been removing some of the Southern whites most likely to accept change in the status quo. Moreover, Donald Matthews and James Prothro showed that Southern counties with highly educated white populations had the lowest levels of black voter registration in the early 1960s, so it will not do to jump to conclusions.

And it surely didn't help that migration tended to remove the best-educated Southern blacks. Between 1955 and 1960, at all educational levels, more blacks left the South than came to it, but as education went up, so did the likelihood of leaving. Add the race and education effects to those of age and sex and you find some extraordinary rates of out-migration: among young, black, male college graduates, net out-migration during the last half of the 1950s alone was over 14 percent—one in seven. Combined with the loss of poorly educated whites, this loss of well-educated blacks exacerbated the already gross differences in educational levels between blacks and whites, as well as taking those most likely to provide leadership for the black community and most likely to confound the stereotypes of prejudiced whites.

Nevertheless, it may be that, like the original frontier, the cities of the North provided a safety valve for Southern society. Presumably the blacks who left included those who most resented and were most inclined to challenge the strictures of Jim Crow. And certainly migration reduced black population percentages in many counties, presumably thereby reducing the well-documented white paranoia that seems to set in when the black population first approaches, then exceeds, 50 percent. In other words, migration may well have contributed in the short run to what whites liked to call "racial harmony"—meaning white peace of mind and black acquiescence in subordination.

And, in the long run, it may have helped Southerners prepare for change. The magnitude of the migration meant that few Southerners did not have kinfolks, or at least former neighbors, living outside the South. Visits and return migration ensured that most would hear firsthand how things were done elsewhere. One student of these matters, Harold Grasmick, has written of the South's "traditional value orientation." For Southern whites this orientation included white supremacist attitudes; for blacks it included fatalism; for both groups it was eroded by exposure to urban, industrial society, whether actual, through travel or residence, or vicarious, through education and the mass media. Like the two World Wars, like the radio and, later, television, migration chipped away at the South's insularity, exposed it to cultural modernity, and paved the way for change.

What happened to those who left? Well, it is hard to generalize, but let me mention a few different types of adaptation, and say a word or two about each.

I have sometimes said that we ought to map the South by identifying Southerners and saying that the South is where they come from. But if we do that, we find patches of the South—well, in the North.

Southern migrants haven't settled at random across the Northern and Western countryside. Whether black or white, these migrants followed the classic patterns of U.S. immigration. In particular, they often displayed what demographers call "chain migration," sending back for friends, neighbors, and kinfolk to join them in these enclaves. Most migrants, in other words, settled where other Southerners already were. The result has been Southern neighborhoods (mostly poor) implanted in Northern cities and Western fields, more or less segregated from the surrounding society—and from each other, on racial lines—with their own churches, taverns, radio stations, and other institutions. Some of these are substantial communities.

Fred Powledge points out, for example, that the Brooklyn ghetto of Bedford-Stuyvesant has a population the size of Fort Worth, or Toledo, and that most of its residents are Southerners by culture and descent, if not by birth. "Some buy their greens and hogmeat from a place called the North Carolina Country Store, which operates its own small-scale truck line to bring the delicacies of home." Smaller versions of Bedford-Stuyvesant can of course be found in every large Northern and Midwestern city.

Black migrants usually had no choice in the matter of segregation: it was imposed on them. But many might have chosen it anyway, as poor white migrants from the South did in the Uptown area of Chicago, for instance. Or in Ypsilanti, where white Southern auto workers are concentrated. Or in Bakersfield, a substantial country-music center thanks to the children and grandchildren of Okies and Arkies. Or in the small town in Washington State where expatriate North Carolina timber-workers, now in their third generation outside the South, maintain a Southern Baptist church and the Tar Heel Cafe.

Both the Dark Ghettos and the Little Harlans of Northern cities were, and still are, characterized by poverty, transience, and unattached young males, with the usual resulting problems of crime and family disorganization. Both provoked pretty much the same alarm in their new neighbors as earlier immigrant neighborhoods had. By and large, however, the pluralistic cities of the Northeast and Midwest have accommodated these transplanted Southern communities as just additional tiles in the ethnic mosaic. For the most part the similarities between "Little Harlan" and, say, "Little Italy" are more striking than the differences.

But one important difference is that return to the homeplace has been easy for Southern migrants, and for many it has been frequent. After all, the Chickenbone Special ran both ways. The Dixie Highway goes south from Ohio as well as north to it. Powledge points out that thousands return to the South from Bedford-Stuyvesant every third or fourth weekend, maintaining family ties "down home." Because the break with the "Old Country" is not as complete, the pressure to assimilate may not be as great, and many transplanted Southerners are only temporary Northerners, if that. In this respect, the Southern diaspora resembles more the Mexican invasion of our day than the European immigration of the late nineteenth and early twentieth centuries.

This more or less successful transplantation of Southern communities to Northern settings, maintaining down-home ways and some measure of ties to the homeplace, is characteristic of people who left the South for economic reasons. These are the men and women who sing the lonesome, homesick songs of country music. In the fifties Mel Tillis wrote in "Detroit City" that he "dreamed about those cotton fields and home." In the seventies, it was Billy Joe Shaver, stuck in Los Angeles with a "bag full of yesterdays and a mind full of doughnut holes," who missed the South. Much has changed in country music and in life, but one constant in both is the homesick Southerner, with "Dixie on my mind." And before you conclude that this genre depends on contrasts between a small-town or rural South and a big-city North, ponder the fact that Emmy Lou Harris and Bobby Bare have recorded homesick songs about Atlanta and Houston, respectively.

At the other end of the spectrum from the economic migrants who sing these songs are what we might call "life-style" migrants, people who left the South because they wanted something they could not find there—and not just a job. We tend to hear from expatriates who miss the South, but we should not forget that there are other migrants we don't hear much from, because they don't miss it: they are happy where they are. In the extreme case, we are looking at *ex*-Southerners, people who have been assimilated to their new host culture, who identify with it, and minimize—maybe even deny—their origins.

Willie Brown, for instance, speaker of the California House and adviser to Jesse Jackson, left his Texas birthplace as a boy and didn't return for thirty years. There is probably no sense in which he is any longer a Texan: he is a Californian now, and doing very well in that capacity. I think also of some boys I grew up with in East Tennessee. One is gay, and last I heard was in Los Angeles, doing what I don't know. Another is a Unitarian minister in New England. A third teaches French at an Ivy League university. And another lives in Marin County, where he is heavily into the Northern California life-style. As far as I know, none of these men has looked back since he left the South. Regional identity just is not particularly important to my friends. Day to day, they "pass," and many who know them don't even think of them as Southerners. Perhaps they aren't, anymore.

Ironically, though, these men are contributing to the persistence of regional differences. By doing their things in places where those things are already done, they make New England look more like my idea of New England, California more like my idea of California. And by leaving the South they have contributed to the relative absence in the region of gay, sushi-eating, Unitarian post-structuralists.

Life-style migrants like these may be able to forget that they are from the South. Maybe the economic migrants in the Southern enclaves of the North can sometimes forget that they are not *in* the South. Those are the two extremes of adaptation: assimilation and isolation. But many migrants manage to move more or less in the Northern mainstream without minimizing their Southern origins. (Some, indeed, trade on those origins, a point I will come back to in a minute.) What about them?

Well, like many of the poor Southerners in the black and white ghettos, some think of their Northern sojourn as temporary or provisional. Joan Didion has written that the young Southerners she knew in New York were different from other young provincials in that they always seemed to know the airplane schedules. For expatriates like this, "home" remains the South, even if they never actually move back to it. But others have made more permanent homes outside the South. How have they managed to remain Southerners? What does that mean?

I want to talk about three more kinds of Southern expatriates, without claiming that they exhaust the possibilities. First are migrants who present themselves as exceptions to their hosts' negative stereotypes of Southerners, but do not seriously challenge those stereotypes (and possibly even reinforce them). Second are migrants who exploit the positive aspects of Northerners' stereotypes (there are some positive aspects) and present themselves as somehow "typical" Southerners, in those positive senses. Finally, rare but perhaps most interesting, are migrants who use their Southernness as a vantage point from which to criticize the "American"—that is, Northern—society they see around them, and either don't care what others think of them or are very cunning about what they are up to. You will probably recognize all three types when I give you some examples.

The first, for instance, is exemplified by those émigrés who have always been a feature of Northern cities, men and women in dissent from one or another of the South's many orthodoxies. Some have been almost exiles, alienated by what W. J. Cash called the South's "savage ideal" of conformity. Indeed, some have been virtual refugees, driven from the South for their unpopular opinions. These Southern expatriates have produced an extensive literature, much of it cataloged in Fred Hobson's fine book, *Tell about the South.* It may be that the decline of Southern orthodoxy will mean the end of this genre. It is probably significant, for instance, that Willie Morris came back to the South without recanting the opinions expressed in *North toward Home,* and it is hard for me to imagine that book being written in the 1990s.

Still, for over a century the Southern émigré was a fixture on the Northern scene, and the type is not unknown today. Surely the émigré's criticism of the South has seldom been calculated or hypocritical; Lord knows there has been plenty to criticize in Southern life. But it is also true that dissent from Southern orthodoxy has been as popular in some Northern circles as it has been unpopular in the South, and some of these dissenters have done well by their opinions. At the very least, criticizing the South has been a good way to get dinner invitations.

The Southern poet and novelist George Garrett has written about this phenomenon, speaking of "those like Willie Morris and Larry King and Tom Wicker and Marshal Frady, who have gone North and have actively joined the establishment, establishing their own impeccable liberal credentials by denouncing Southern life . . . while managing to preserve something of the Southern style for themselves." (You may gather that Garrett does not approve of these folks, but you don't have to agree with his sentiment to accept his description of what is going on.) Garrett observes that "the influence of these Southerners-in-residence can be pernicious in the sense that, since they are mostly journalists . . . and since they almost always say precisely what is expected of them, they are regarded as experts and are influential not so much in passing judgment on Southern actions and events as in enforcing conventional judgments already passed by others."

These dissenters do not shed their Southern identity—indeed, their status as dissenters depends on their maintaining it. But, as Garrett observes, they may help to perpetuate their Northern hosts' negative images of Southerners. They just present themselves as exceptions to the rule—as, indeed, they are.

Compare these Southerners to another set of expatriates, whose success, when they find it, depends precisely on being perceived as somehow "typical." Sam Ervin, Burt Reynolds, Tallulah Bankhead, William Faulkner, and Perle Mesta ("the hostess with the mostest") have in common that they are among the many Southern politicians, movie stars, literary men, and society hostesses who in one way or another have exploited Northerners' images of Southerners as courtly, hospitable, "authentic," romantic, or at any rate somehow interesting. (To use your origins this way requires that people recognize them, by the way, which is why I am talking here mostly about white Southerners. I don't think most Northern whites yet see Southern blacks as black Southerners.)

I've written a small book, *Southern Folk, Plain & Fancy,* on this subject, so I won't dwell on it here, but let me give a few examples, drawing on the history of Southerners in Gilded Age society. Someone needs to write a book about this, by the way—sort of a Southern version of *Our Crowd.* A book by William Stadiem called *A Class by Themselves* makes a start, but one of Stadiem's earlier books was *Marilyn Monroe Confidential,* so you can probably tell that he's not too serious about it. Still, let me introduce you to a few of the characters in that book.

Consider, for example, Richard T. Wilson, a former traveling salesman from Loudon, Tennessee, who made a shady wartime fortune selling purloined Confederate supplies in Europe. Wilson was a tall, handsome man with elegant manners, and after the war he moved with such success in New York society that his family became known as the "marrying Wilsons." His son married Carrie

Astor, and his daughters married a Goelet millionaire, a Vanderbilt, and "Mungo" Herbert, the brother of the Earl of Pembroke.

Southern gentility was also played to advantage by the Smith sisters of Mobile: one married an English nobleman, another a Cuban sugar planter, and a third, Alva, married William K. Vanderbilt and then (after a scandalous divorce) Perry Belmont. (A fourth sister chucked the whole business, remained unmarried, and became an ardent feminist—as did Alva, for that matter.)

Other Mobile girls snared the widowed Commodore Vanderbilt and the Duke of Manchester, but when it came to raising export-quality Southern belles, Mobile couldn't hold a candle to Charlottesville, Virginia. The five Langhorne sisters of Charlottesville made at least eight socially noteworthy matches: one to an English merchant banker, later Lord Brand; another to the society painter Charles Dana Gibson (she became the prototype of the "Gibson Girl"). Still another left her first husband to run away with a star of silent cowboy movies and former Yale All-American football player. And Nancy Langhorne married one of the English Astors, to become Lady Astor and sit in Parliament.

Charlottesville also produced Amalie Rives, a beautiful, best-selling novelist who married an Astor grandson, and Beatrice Ashley, who married his brother. As a fetching seventeen-year-old, Beatrice sang "I'm a Naughty Girl" in the New York production of *The Greek Slave.* Amalie's husband was a dashing big-game hunter known among his friends as "the General" because of his recurring conviction that he was Napoleon Bonaparte. He liked wild duck and ice cream for breakfast, wore leather pajamas, was charged with "at least one" murder, and was often in residence at New York's Bloomingdale Asylum. After Amalie divorced him, she married Prince Pierre Troubetzkoy, to whom she'd been introduced by her friend Oscar Wilde.

I must also mention Princess Amalie's friend Princess Alice—née Heine, of New Orleans. Alice married the Duc de Richelieu in 1875; after his death she married Prince Albert of Monaco, edging out Philadelphia's Grace Kelly by nearly seventy years. But all this business of Southern American Princesses is starting to sound like a gossip column, so I'll get back to the point, which is that some Southerners outside the South have played the image of Southern gentility (perhaps especially the mystique of the Southern belle) for all it's worth.

For them, the advantages of being "typically" Southern have outweighed the disadvantages. Being Southern also explained inconvenient facts like why the mother of the Smith sisters had kept a boardinghouse, or why the Langhorne girls' father had been a tobacco auctioneer: those distinguished families had been impoverished by The War.

But if you're not Lady Astor, acting out somebody else's idea of what Southerners are all about can be a risky business. There can be an element of Samboism, of pully-woolly, in all this. It may be relevant that several of the women I have mentioned were early activists in the cause of women's rights. ("I married beneath myself," Lady Astor said once. "All women do.")

You also risk becoming something of a house Southerner, subject to dismissal, as Ward McAllister learned the hard way. A Southerner who had become Mrs. Astor's confidant and (in his own view at least) an arbiter of Newport society, McAllister published *Society As I Have Found It* in 1890. Some of his Yankee friends believed, with reason, that McAllister portrayed them as parvenus who needed his social guidance. Thereafter Mrs. Astor was not at home when he called, and Stuyvesant Fish announced, "McAllister is a discharged servant. That is all."

I gather something of the sort happened to Truman Capote when excerpts from *Answered Prayers* were published. Unlike McAllister, though, Capote may have known what he was doing. Capote may have been an example of the last type of Southern expatriate I want to mention: those who have compared the North unfavorably to the South and quite consciously set out to trash it. A good many talented Southerners—from the Vanderbilt Agrarians to Tom Wolfe (the New Journalist, not the Tar Heel novelist)—have had that response to living in the North, and the literature of American social criticism is richer for it. Oddly, some of them are richer for it, too. In an essay in a book called *Whistling Dixie* I've tried to figure out how they get away with it. But let me close here with a paradox.

Outside the South, the South is surviving nicely. It retains its vitality in the hearts of most Southern expatriates, and it is also alive and well in the heads of non-Southerners, whose ideas of the South, however out-of-date or otherwise inaccurate, are part of the environment to which Southerners must adapt. When I talk to audiences of non-Southerners about the South, they don't doubt that there is something there to talk about. After all, they sometimes run across "Hee Haw" on television.

Lately, however, I have run into more skepticism about the South's vitality *in the South*. It seems that some Southerners have begun to wonder whether "the South" still exists, in any sense that requires us to take it seriously. This question seems to be especially common in places like piedmont North Carolina that have experienced considerable in-migration lately. We need to look at that in-migration, at how Northern migrants are adapting to the South, at some of the problems their presence poses, and at the implications of their presence for the future—even for the future existence—of the South. But that is a large enough subject to need another essay.

Refugees and Returnees
The Southern In-Gathering

■ In the recent past—that is, since 1877 or so—newcomers to the South have been overshadowed both numerically and symbolically by Southerners leaving the region. For the first two-thirds of this century the South each year lost far more migrants than it received. In the 1930s and 1940s some folks even took to calling the South the "seedbed of the nation," an upbeat way of saying that we were shipping our young seedlings in large numbers to other parts of the country.

Since about 1960, though, more whites have been moving to the South than leaving it, and since the early 1970s, the same has been true for blacks. What's going on here?

We need to recognize that this turnaround largely reflects the fact that Southerners are not leaving the way they used to. The decrease in out-migration needs just as much explaining as the increase in *in*-migration. I suspect, however, that Southerners have stopped leaving for pretty much the same reasons that outsiders have started coming. By 1978, for example, North Carolina State University was reporting that more than half its science and engineering graduates were staying in the state; in the 1960s, fewer than a quarter had stayed. Those graduates were kept at home by the same economic changes that have lured outside talent to North Carolina.

Migration to the South is an old story, of course. One of the untold sagas of American history is that of the Great Trek of the Scotch-Irish down the Wilderness Trail to settle the Southern interior in the eighteenth century. Eli Whitney was only one of many Yankee immigrants whose activities affected the Old South profoundly. After Reconstruction, many of the two hundred thousand occupation troops stayed on; twenty to fifty thousand of them just bought or leased farms at depressed prices and blended into the local population. (They must have descendants, although I've never met anybody who admits being one.) In an unpublished paper, Robert Freymeyer has pointed to a number of similarities between the post–Civil War influx of Yankees and the current, what we might call "Sunbelt," migration. It seems the carpetbaggers were the yuppies of their day.

Yes, there has always been migration to the South. But since World War II it has begun to be measured in millions, and, to repeat, now it exceeds migration *from* the region. In many Southern towns, including mine, well over half the residents are people who weren't there ten years ago. Who are they? Why are they coming? And why now?

If the image that comes to your mind is one of fast-talking Yankees swarming across the Southland, let me reassure you. When we take the statistics apart, the picture that emerges is less alarming. To begin with, many newcomers to Southern towns and cities come from other places in the same state, especially from rural areas, many of which are still losing population. And those who come from other states are likely to come from adjoining ones. A quarter of the residents of Georgia, for example, are now out-of-staters, but most of that quarter were born in Alabama, Tennessee, North or South Carolina, or Florida. Non-Southern interstate migrants outnumber Southern ones only in Florida, Kentucky, Oklahoma, Texas, and Virginia—all, you will notice, on the edge of the region.

Massive in-migration by itself can be unsettling and socially disruptive—or, for that matter, exciting and challenging—but it surely makes a difference where the migrants are coming from, and my point is that many of those moving to fast-growing parts of the South are Southerners. Much of the rural South is still losing population, as I said, but increasingly it is losing it to Southern cities, not Northern or Western ones. That many graduates of my East Tennessee high school leave East Tennessee is nothing new. What is new is that many now go to Charlotte, or Atlanta, or Nashville, rather than Akron, Dayton, or Cincinnati.

Still, in just ten years during the late 1970s and early 1980s over four million people moved to the South. Nobody is yet calling the North the seedbed of the nation, but that is a lot of folks. Who are *they*?

Well, many of them are not Yankees either. They are "return migrants," Southerners who left the South to seek jobs or education, or to escape discrimination, now coming home. As a matter of fact, I'm one of them: the 1960 census showed me as a migrant from Tennessee to the Northeast; the 1970 census as a migrant from the Northeast to North Carolina. All I did was go to college outside the South.

Return migration is an especially important component of in-migration for blacks and for older whites. In the late 1970s, Ronald Snow looked at 502 black migrants to Mississippi from the Northeast and Midwest. He found that 272—54 percent—had previously held Mississippi driver's licenses or had social security cards issued there. Undoubtedly many of the others had left Mississippi as children, or had been raised in other Southern states. The equivalent figure for white migrants to Mississippi was considerably lower (12 percent), but it was higher among older people. That finding was reinforced by some research on retirement migration by Charles Longino and Jeanne Biggar, who found that most retirees from the North to Southern states (other than Florida) are returning Southerners.

Return migrants, of course, are not the same as Southerners who have never left. Among blacks, for instance, they tend to have higher levels of education and achievement than stay-at-home Southern blacks (and higher levels than Northern blacks, for that matter), one reason that black return migrants have become important figures in many Southern communities and are increasingly an important factor in Southern politics. White return migrants, too, have been affected by their time outside the South; some survey research shows, for example, that they are more likely than whites who have never lived outside the South to think of themselves often as Southerners, even though their attitudes are less typically "Southern" in some respects. But the important point here is just that a good many of the people now coming to the South are the same ones who left in previous decades.

Still, there are some genuine Northerners among the new crowd of migrants. One out of every eight people in the sixteen-state "Census South" now hails from some other region. What do we know about them?

Funny thing: not much. Since foundations and government agencies have tended to see Southern migration to the North as a problem, we have some studies of Southern migrant communities. But nobody seems to be worried about the effects of Northern migration to the South.

Not that it's not a problem. There is, to begin with, a problem of etiquette. What do you say when someone tells you he comes from Boston? "I'm sorry"? Over a hundred years ago, Joseph Baldwin suggested that the polite response is to pretend that you never heard of Boston, but that probably won't work anymore. They've been in the papers a lot.

In some parts of the South, of course, the problem no longer arises, because nearly everyone comes from Boston, or someplace like it. A Chapel Hill friend of mine was present when two of his new neighbors discovered they were both from New Jersey. "Oh, yeah?" one of them said, "Which exit?" Flannery O'Connor used to tell the story of an Atlanta real estate agent who told a migrant client, "You'll like this neighborhood. There's not a Southerner for miles." Miss O'Connor professed to take some comfort from the fact that at least Southerners can be recognized when we do occur.

These stories bear on one of the factors that affect both the probable extent of migrants' influence on Southern culture and the likelihood that cultural differences will lead to conflict. Both are decreased if migrants keep to themselves, and to some extent they do. I will come back to that. When two different cultures do come into contact, however, some degree of misunderstanding is almost inevitable. (That Southern culture *is* different is something most migrants —in either direction—are ready to tell us from their own experience.) Sometimes cultural differences lead to conflict, more or less serious.

Now, this sort of conflict is the exception, not the rule. (I may give it more emphasis than it deserves, just because it is more interesting than mutual forbearance and respect.) Conflict is not always a bad thing, either. It can produce needed change, as it has from time to time in the South. And, perhaps because this particular conflict is low-level and chronic rather than lethal and acute, sometimes it can even be amusing.

One more thing. It seems to be almost impossible for a Southerner to talk about sectional conflict without being accused of "still fighting the Civil War." Do I have to say that talking about conflict isn't the same thing as advocating it? Live and Let Live is my motto. I always tell Yankees to bring the spoons back and we'll forget all about it.

Anyway, in any community the amount of cultural conflict between migrants and natives will necessarily depend in part on their numbers. In much of the South there is no conflict because there are so few migrants that they have no choice but assimilation or social isolation. That's the situation Lewis Killian described in *White Southerners,* suggesting that, as late as 1970, the impact of migrants was negligible in most of the South.

Where migrants are present in truly overwhelming numbers, there is also no problem: they become the community that the *natives* must fit into. That's the case in South Florida, for example, and in what a friend of mine calls "Occupied Virginia." Chapel Hill may be headed that way. A North Carolina boy complained recently in our student paper that people are always coming up to him and saying that they love to hear him talk. "Let's get one thing straight," he wrote. "We are in the South. Therefore, I do not have the accent." What really annoyed this lad, though, was "having to defend my region when I'm still in it." He'd better get used to it. In places like the Carolina piedmont, we may have a rough patch to get through. At the very least, natives are going to grumble about folks who don't know grits from granola. Even in liberal Chapel Hill, for instance, I've heard people refer to Northerners as "'rhoids"—short for hemorrhoids, from a rude joke with the punch line: "If they come down and stay down they're a pain in the ass." (There's a great scene in the movie *Sharky's Machine.* Burt Reynolds, an Atlanta vice cop, confronts Vittorio Gassman, the white-slaver villain, and says, "I'm gonna pull the chain on you, pal, and you want to know why? Because you're [mess]in' up my city. Because you're walkin' all over people like you own 'em. And you want to know the worst part? Because you're from out-of-state.")

In parts of the South where economic activity has attracted assertive newcomers who don't look Southern, culturally, and don't want to, and where there are enough of them to make a fuss, but not yet so many that they have rearranged things entirely to their satisfaction, any number of issues can take on an us-against-them aspect, pitting locals against out-of-state Northerners.

Consider the "Super Tuesday" primaries of 1988. In North Carolina, at least, the division in both parties was remarkably clear. In the Democratic primary, while Albert Gore and the Redeemers were battling it out with Jesse Jackson and the freedmen, the Dukakis fifth column snuck off with a substantial chunk of votes—most of them, I would bet, from migrants. Ironically, across the aisle, it was Pat Robertson's Southern legion, black and white together, who were seen as the pushy outsiders. In our neck of the woods, at least, the principal champions of the old Republican ways were retirees from Bush and Dole country.

Different states have had different experiences, but in North Carolina we've had conflicts over the content of public-school textbooks, over the display of Confederate symbols, over Christmas decorations on public buildings, over school prayer and corporal punishment, over the Equal Rights Amendment and liquor by the drink. None of these has turned into a straight-up migrants versus natives fight: in each case, liberal newcomers have found local allies, and some migrants have been conspicuous in the resistance. But you'd have to be pretty obtuse to miss the sectional undertones in the letters-to-the-editor columns.

My favorite example—I've used it before, but you can't overdo a good story—comes from Hillsborough, North Carolina, down the road from us. Formerly, if a Hillsborough squirrel was giving you trouble you called the police, who sent an officer over with a .22 to take care of the problem. The theory was that this beat having private citizens shoot squirrels on their own, and I guess it does. This arrangement was working just fine until a newcomer went to the town council and objected to it. Now squirrel shooting in Hillsborough is hedged about with all manner of restrictions. Owners of all adjoining property have to be notified before a squirrel is shot, the person who wants it must be given a leaflet describing kinder, gentler methods of getting rid of the creature, and it's a good deal harder to kill a squirrel than to get an abortion. Does this make sense?

I don't know. Maybe it's a joke. When cultures collide, humor is one of the things that translates least well. A while back the *New York Times* interviewed a woman from the Bronx who was teaching at the University of Texas. She said: "Texans have no sense of humor. They can't tell when they're being kidded." When I was in school in Boston, we Southerners used to say the same thing about Ivy Leaguers.

It helps to be bilingual. One twenty-year veteran of doing business in North Carolina remembers telling the president of a New

England company that he, the veteran, would be useful to them because he spoke both Yankee and Southern. He says: "They just laughed and brought down their own Yankee personnel man. He lasted less than a year; he didn't know the 'language' nor understand Southern ways." That's the kind of low-level problem we're having in the South. Problems of misunderstanding, not a fundamental conflict of interest.

But they are real problems. One of the great remaining regional differences has to do with how criticism is understood. I suggest, at the risk of overstating, that some Northerners—New Yorkers, in particular—believe that criticizing others is a God-given right, enjoy doing it, and admire people who do it well. Southerners, on the other hand, tend to see it as bad manners. When Northerners criticize, in other words, they do it forthrightly, and they may not mean any harm. Sometimes they're even trying to help. When Southerners criticize, they either do it very indirectly, or they *intend* to give offense. One businessman from Ohio, now in Georgia, complained about this. He told *U.S. News and World Report,* "If [Southerners] think a guy is an SOB they'll apologize before they say it. I wish they'd call it like they see it." But, as someone said once, Southerners will be polite until they're angry enough to kill you.

This difference in manners can work to the disadvantage of Northerners who come South. Some don't see any reason to stop complaining about things they don't like, and sometimes what they don't like is Southerners. But Southerners don't appreciate criticism, even if it's well-founded. Perhaps especially if it's well-founded.

Now don't misunderstand *me.* Some migrants don't find anything to complain about. No doubt some keep their opinions to themselves. But others don't hesitate to express their criticism. Sometimes they do it in the pages of Southern newspapers. I have been collecting examples for a few years, and I have some examples of Southerners' responses, too.

Ironically, what some Northerners criticize us for are the same things that other Northerners admire. Research on regional stereotypes shows that Northerners are likely to see Southerners as slow (sometimes with connotations of stupid), as conservative in a variety of ways, and as friendly and polite. Each of those traits can be evaluated either positively or negatively. (You might ask how somebody can complain that someone else is friendly. I'll get to that.)

By far the most common observation about the South is that people are slower here. Indeed, that seems to be the paramount Southern characteristic in the minds of many Northerners—and Southerners, too, for that matter. One of our students, from Boston, told about his first dealings with a Southern country storekeeper: "He talked so slow, I had to stop and slow myself down to understand him. He kept asking me, 'What's your hurry, son?' "

The *New York Times* quoted the migrant wife of an IBM executive, who has lived in Texas for ten years: "They tell me I talk too fast, I walk too fast. They just want me to slow down, and I can't." An article in the *Jackson Clarion-Ledger* quoted a Princeton psychologist, originally from the Bronx, on his first months as a graduate student at Duke: "I thought I had landed on a different planet. There's a tremendous difference in speed and tolerance for delay. For the first few months, I would want to help waitresses take my food to the table. I mean, I'd try to make people's lips move a little faster."

But this man now thinks Southerners may have the right idea, and so does one of our students, from Pennsylvania. "It's not that rush-rush style like up North," she told the *Daily Tar Heel.* "Even my walk looks different when compared with Southern students. They just take their time to class and get there when they get there. Meanwhile, I'm running the four-minute mile." She envies her Southern friends. "It's the way life should be lived, to enjoy each moment."

The point is that some migrants like what they see as the Southern way. Here's a New Jersey woman, now in North Carolina: "I can appreciate the slowness. It's no wonder more people drop dead from heart attacks up North." And a woman from Ohio, now a Georgia housewife: "You just don't see pushing and shoving here. I hope we don't lose that comfortable, friendly pace."

But others find "that comfortable pace" harder to take. Another student, from Illinois, complained: "I'm a very active person and when I came down here almost all the people I came in contact with gave me the impression of being very lazy. They weren't enthusiastic about much of anything and it was very hard to get them excited." And the *Wall Street Journal* quoted one Yankee businessman on doing business in the South: "You go out to visit these CEOs who talk very slow, very deliberate. 'Well, sir-r-r, we're ve-ry interested in long-term valuuuue.' I want to say, 'Come on, guy, spit it out. Talk. I want to get home by next year.' "

Sometimes the observation that Southerners are slow carries the implication that they are stupid or incompetent, and some migrant observers don't stop at implication. One man, more outspoken than most, wrote *The State,* of Columbia, South Carolina, to assert, "If it were not for us d—n Yankees, South Carolina would be 54th in the nation." And a visiting New York student won no friends in North Carolina when he volunteered to the *Tar Heel,* "You know the trouble with Southerners? They're stupid."

A few Northerners even complain about our friendliness and good manners. Sometimes the problem is just that they're not used to them. As a student from Philadelphia told the *Tar Heel,* when he first came to North Carolina, "Everybody was almost too friendly to me. I didn't know how to react to it." Another Northern student agreed: "It ain't easy for a boy from the Bronx to be yessired by cops and cashiers and smiled at by total strangers." He was especially nonplussed by a convenience store clerk who "thanked me with an earnestness that would have been excessive if I had offered to donate a kidney to her sickly grandmother."

True, once they get used to this, most Northern migrants seem to like it. As a Connecticut housewife recently arrived in Roswell, Georgia, told *U.S. News and World Report,* "In a store here you'll encounter 'Y'all come back now.' You walk out feeling good, and you want to come back."

But some migrants observe that manners disguise Southerners' real feelings. (That would sound familiar to the Japanese, by the way. They're polite, too, and they get called "inscrutable.") "Southerners are more apt to say cordial greetings to each other, but that's about all," one student from Illinois told the *Tar Heel.* "It's a very superficial friendliness."

Some Northerners who have figured this out like it anyway. A graduate art student from Boston observed that Southerners' greater friendliness "works two ways." "Some of them are friendly," he said, "and some of them are but really aren't. It's just sort of a politeness." But, he added, "that's fine because it makes things easier anyway."

A Massachusetts businessman, recalling the years he spent in the South, agreed. "A lot of Northerners thought the Southerners' friendliness was phony—saccharine, sugar-coated," he said. "But I didn't care. I'd rather people be nice to me than not nice. If you're going to be the new person in town, the South is a good place to land."

But a few transplanted Northerners seem to be really annoyed by what they see as Southerners' lack of integrity. A department-store executive, relocated to Georgia from Ohio, complained that Southern graciousness "does not come across as politeness but insincerity." And a woman from Philadelphia told a North Carolina journalist, "It's all epitomized by the neo-Southern Bitch. She dresses so damned cute. Who's she think she's fooling? It's all just fluff, and flirt, and manipulation."

The journalist, a Southerner, commented that there was "definitely no fluff" to this woman. "She would not flirt or manipulate: say the wrong thing, and she'd simply rip your ears off." In fact, though, she may have resented the response of Northern men to "fluff, flirt, and manipulation." In general, they like it. A male graduate student from St. Louis, for example, said that Southern college women "don't come on as hard. They're much less aggressive in their relationships." (He also saw them, incidentally, as "more modest" and "pretty chaste in their moral attitudes.") Another student, from Boston, added that Southern coeds are "a lot more refreshing. . . . Down here they have a sweeter image and I like that."

By and large, Northern women tend to like the manners and style of Southern men, too. (Their attitudes may be another matter.) One of our pharmacy students, a New Jersey woman, told the student newspaper, "Southern guys are more polite and they're more apt to do things like hold open doors. I enjoy it. I'm liberated but I'm not going to get pissed off if some guy holds the door open for me." Another Northern woman, from Pennsylvania, agreed, adding, "They don't seem to forget that you're a woman, and that's nice." Well, yes, but sometimes it's not so nice. A little later the same woman was saying that "men are much more chauvinistic in the South."

This leads us to the third, and last, great characteristic of Southerners in the Northern mind. We are seen as more conservative, in nearly all respects.

Once again, some Northerners like that difference. Some even move South to be among people who think the way they think we think. An article in *National Review* urged conservatives to send their children south to college, because "Southern society places a check on the excesses of liberalism." "Professors," the article said, "are constantly reminded (by the gun racks in the pickups and by letters to the editor) that the dominant culture doesn't go for accommodation with the Soviets, buggery in the boys' room, or any other of liberalism's proclivities." (I think the author was joking.)

Readers of *The Nation,* of course, are less enthusiastic about our conservativism. And so is a young woman from Pennsylvania, one of our ever-quotable students, who told the *Tar Heel* a while back: "I thought Pennsylvania was the most backward state until I got down here. . . . Then I found out that when you go into a bar, you can't buy mixed drinks and I thought, 'Oh Yecchh.' "

At least this woman had a complaint something could be done about. Soon afterward, Yankee migrants, with the aid of local quislings, legalized mixed drinks in Chapel Hill. But what could possibly be done to satisfy the woman who wrote the *North Carolina Independent* to demand "change" in North Carolina. "We would all love to preserve traditional culture," she wrote, but "change is inevitable. Without change there is no progress. . . . Change means improvement, but if people are afraid of change they cling to old ways, old dreams, and blame their misfortunes on [those] who try to come in and improve things." "Look at this place," she went on, "it definitely needs outsiders with some determination and guts to improve it." "Wake up North Carolina," she concludes, "you definitely need outside help!"

The bizarre thing about this letter is that the woman never said what sort of change she had in mind. She began with the remark that "I have lived all over the United States and have come to a sad conclusion: North Carolinians are one of the unfriendliest groups of people I've ever met"—and I must say that by the time I finished her letter I did feel distinctly unfriendly.

As a rule, our survey research shows that North Carolinians don't really dislike Northerners; most, for example, say they wouldn't mind if their daughter married one, as long as she didn't move up North. But a lot of otherwise easygoing Southerners seem to be driven to frenzy by Northern missionaries. The *Independent* published another letter, soon after this woman's. It began: "I rarely find it necessary to respond to the Yankee ignorance we've all grown accustomed to in North Carolina." But it continued: "Take your show on the road. Better yet, take your show, your family and your Yankee friends back to the land you ruined and then fled."

As a matter of fact, the man who wrote this wasn't replying to the lady who wanted us to change: he was responding to someone who had suggested that the Atlantic Coast Conference was inferior to the Big Ten. But the characteristic Southern response to criticism does seem to be "love it or leave it."

When a group of expatriates in Austin formed a club called the Damn Yankees and began to gather at each others' houses to eat kielbasa, tsimmes, challah, and chopped liver, and to go together to movies like *New York, New York,* nobody minded that. But when they wrote the *Daily Texan* to complain about Texas-sized cockroaches, weather forecasters who stand in front of the northeastern part of the map, and the absence of bagels and taxis in Austin, a counter-organization sprang into being called "Go Badmouth the Austin Community to Your Asinine Northern Companions." (The acronym is GO BAC YANCS.)

Southerners' response to criticism, of course, gives their critics something else to criticize. But enough war stories. What's the point? Just that some measure of conflict does accompany the arrival of Northern migrants in Southern communities, and that we can expect more of it as more of them come to live among us. As the lady says, without change there's no progress. But I say unto you that without conflict there's not much change—perhaps especially in the South.

So the question of how much impact migrants will have on Southern culture is related to the question of how much conflict there will be. And I want to look at four factors that will affect the answers to those questions. I mentioned one already, when I said that the impact of migrants and the likelihood of conflict are both reduced to the extent that migrants avoid association with the natives. Both conflict and impact are also minimized if migrants see themselves as just passing through, if they are already culturally "Southern" when they come, or if they are quickly assimilated to Southern culture. All four factors—isolation, transience, self-selection, assimilation—are at work to some extent. Let me say a word about each of them.

As I said, one out of eight residents of the "Census South" comes from what we can call, for short, the North. But they are not scattered at random around the South. They are concentrated on the periphery of the region—in Texas, Florida, the Ozarks, and northern Virginia—and in a relatively small number of fast-growing cities elsewhere. Nearly two million people moved to the South between 1980 and 1985, but four Southern states (Alabama, Mississippi, Kentucky, and West Virginia) lost population. In that same period, over three million new jobs—more than half the nation's total—were created in the South. But more than half of those were in Texas and Florida, and another third were in a mere eleven urban areas in other states. All told, that is, nine out of ten of the South's new jobs in the early 1980s were in Texas, Florida, and a dozen cities elsewhere. Mississippi added fewer than nine thousand jobs, and West Virginia lost fifty thousand.

So a good deal of the South's population and even more of its acreage have not seen much in the way of Northern migration. In most Southern states, the proportion of residents who are Northern-born is well below the regional average—more like 5 to 10 percent. Florida and Texas blow the curve. And for every Southern community that feels itself overwhelmed by the influence of Northern migrants, there are a half-dozen towns like Lumberton, North Carolina. Karen Blu, a Yankee anthropologist, complained in her book about Lumberton that she had to drive all the way to Fayetteville to get the Sunday *New York Times.*

Like Southerners in the North, Northern migrants have had an impact on the politics of the cities and states where they congregate. In our system, political clout is magnified by geographical concentration. You say transplanted Southerners drove up the George Wallace vote in Michigan in 1968? Well, as I mentioned, transplanted Northerners voted for Dukakis on Super Tuesday twenty years later—and the two Southern states whose primaries he carried, you'll recall, were Florida and Texas.

But the cultural impact of Northern migrants is reduced when they settle in enclaves and avoid the natives. And many do. I can't resist another story. My friends and neighbors in Chapel Hill have solved the problem of racial imbalance in our public schools with an ambitious program of busing. But I'm told that some schools in our system are still less desirable than others, because they contain large numbers of indigenous white North Carolinians. No one has yet proposed busing as a solution. There seems to be some question about whether white Southerners are ready for equality.

Although Northern migrants to the South resemble Southern expatriates in the North in their tendency to segregate themselves, poor black and white Southerners in their Northern ghettos know that *they* are the outsiders. As late as the fifties, residents of migrant enclaves in the South knew that, too. Such cantonments were oddities in the South. Martin Duberman records the reaction of a visitor to one of them, Black Mountain College, near Asheville. In its "hillbilly setting, in the Southern Baptist Convention country of the Tarheel State," this visitor felt, it was "a little like finding the remnants of an advanced civilization in the midst of a jungle."

The proliferation of such compounds required a critical mass of migrants. But now we have it, and there are so many enclaves, so new and so conspicuous, so often found in those parts of the South that visitors are likely to see and where even native commentators are likely to live, that they get more than their share of attention. I repeat: we should not forget that most of the South is not like that.

But we do need some serious studies of how these places work. It seems to me that they work pretty much the way the "Little Harlans" of the Midwest work. The "Little Greenwiches" springing up around many of our cities shield migrants from a bewildering new culture, provide them with their own churches, restaurants, social clubs, and other institutions, and allow them sometimes almost to forget that they have left home.

Like the residents of hillbilly ghettos, those who live in these new settlements tend to be economic migrants, not "life-style" migrants. What they want from their new region is jobs, not much else. Many denizens of corporate America transferred to Southern outposts want communities where they might as well be in the suburbs of Minneapolis or Hartford, and (aside from the weather and the need to deal with local service workers) that is pretty much what they get. Enclaves make the transition less wrenching, because there is little transition involved.

As a rule, something like two-thirds of newcomers to American communities report that they like the new place better than the old. So our local boosters were dismayed to discover in 1982 that only half the newcomers to the Raleigh-Durham area preferred it to where they came from. But they should not have been surprised. Our area's growth means that many newcomers are now economic migrants, people assigned to North Carolina rather than those who have chosen voluntarily to come. And the nature of our area's growth, in consequence, has made it less and less distinguishable from San Jose or suburban New York. Forty percent of the newcomers said the Research Triangle area was similar to where they came from: why say you prefer one to the other when you can't tell the difference?

The same thing is happening in other fast-growing Southern cities. Listen to Joel Kweskin, a Charlotte advertising executive. Over a platter of Szechuan chicken in a Charlotte restaurant he told the *New York Times:* "I've always lived in the New York area and I never thought I could live anywhere else. But . . . most of the people you meet here in business are from the North or Midwest. Our lifestyle hasn't changed that much from when we were living in Fair Lawn, N.J." Precisely.

Even the architecture of these ghettos is designed to be like what migrants are used to. Never mind that an Atlanta realtor told the *Wall Street Journal* about the "Gone with the Wind atmosphere" of the "mega-homes" on that city's outskirts. What that means is that some have columns (most likely of extruded aluminum, as advertised in *Southern Living*); behind those columns are three-car garages, master-bedroom suites with sunken tubs and wet bars, two-story marble-floored atriums—all the clichés that identify the homes of the nouveaux riches from sea to shining sea in the waning years of the twentieth century.

These ostentatious dwellings are for transplanted top executives, and for the local realtors who are cleaning up on all this activity. They are not the only kind of migrant housing, of course. There are townhouse apartments and condo complexes for singles and "dinks" (double income, no kids). There are blocks of cookie-cutter houses with standardized amenities for the families of professionals and middle managers. But all have in common that they are fungible, easily assessed by realtors for sale to succeeding generations of transferred professionals and executives. This is not the kind of thing you buy with the thought of bequeathing it to your eldest son.

And that leads me to my second reason for suggesting that migration's cultural impact will not be as great as you might suppose from the numbers: many migrants are just passing through. Like many Southern migrants in the North, some are essentially *Gastarbeiter*—guest workers—keeping their ties to home, and planning to return in time. I'm waiting for the music:

> Last night I went to sleep in DeKalb County
> And I dreamed about those traffic jams back home.

Can't you hear it?

You've probably run into the statistic that one American family in five moves each year. What that does not say is that it is pretty much the same family year after year. Economic development in the South means that we are on the circuit now. So a lot of these folks don't plan to be here long, and are not likely to invest much of themselves in the community. People who know that they'll be moving on shortly are less concerned to change what they don't like.

The case is probably similar for most of the retirees, who also aren't planning to be around very long. Anyway, they probably like it. That's why they came in the first place.

Which illustrates the third of the four factors I mentioned: self-selection. Migrants from the North are not just a cross section of the Northern population. We get life-style migrants as well as economic migrants. Some people, in other words, come to the South because they like it the way it is.

Some time back, on a flight to Nashville, I sat next to a young woman with a baby. The baby was drooling on my sleeve, so I struck up a conversation. It turned out the woman and her husband were from New Jersey, born Roman Catholics, but converts to the Church of God. They had moved to Nashville to escape what they saw as the decadence of New Jersey and to raise their children in a "Christian environment."

Think about that. Here is an example of what we might call "assortative migration." If all the God-fearing people of New Jersey check out and head South, the effect will not be to *reduce* regional differences.

That's an extreme case, but a couple of researchers who have looked at survey data on migrants suggest that it is not all that unusual. On the average, migrants to the South already look something like Southerners, culturally. In a few respects, they look more like Southerners than Southerners do. They are even less wild than the natives about taxes and federal programs to help the poor, for example.

So we should not carelessly adopt a mixmaster image of migration effects. That may be an appropriate model for dealing with economic migration, but life-style migration can exaggerate existing regional differences, or even produce them where none existed before. By moving to California and New England, some atypical Southerners have made those places look more like California and New England, and the South more like the South. Just so, some migrants to the South make the South more Southern, not less.

The final factor working to reduce the cultural impact of migration is the South's remarkable capacity to absorb and assimilate migrants. Remember those disappearing federal soldiers. Survey data show that something like 90 percent of native North Carolinians believe that they are living in the best state. Merle Black has analyzed those data, and he shows that newcomers are much less likely to believe that. But migrants who have lived in North Carolina more than five years *look like natives.* Presumably some who don't feel that way leave; others apparently change their minds.

In my book *Southerners,* I looked at survey data from residents of North Carolina who had been born and raised outside the South. Roughly a quarter of them told us they now think of themselves as Southerners. Some had married Southerners, and others just said they had come to "feel closer" to Southerners than to other Americans. I haven't looked at data on the children of migrants, but surely they would show even higher levels of assimilation.

Nearly every recent traveler's account of the South offers an assimilated-migrant story or two. One of the more unusual comes from Joel Garreau's book *The Nine Nations of North America.* Garreau encountered a German industrialist in Jackson who had four flags on the wall of his den: the U.S. flag, the West German flag, the flag of his German hometown—and the flag of the late Confederate States of America. This man still talked about the "Norse" and the "Souse," and he joked that he wasn't enough of a Mississippian yet to own a Luger. But he complained to Garreau about Yankees' stereotypes of the South, and Joel noticed that the German's teenage daughter was "a blow-dried blonde heartbreaker" who looked and sounded like the Ole Miss sorority girl that she almost certainly became.

That's assimilation. Along with isolation, transience, and self-selection, it operates to reduce the clash of cultures that more or less inevitably accompanies contact between a settler population and an established native civilization.

And we can be grateful for anything that does that, because conflict between settlers and natives elsewhere in the world has seldom been amusing, and it is often very ugly. But we are not talking here about the West Bank, or Ulster, or Wounded Knee, or even Quebec. What is going on in the South today just isn't in the same league, and it would obviously be absurd even to suggest that it might get that way. In the South, I confidently predict, conflict will remain the exception rather than the rule. When it does take place it should usually be of low intensity and short duration, and more often than not relatively good-natured—like the examples I gave you, and like the experience of Southern migrants in the North.

The factors I have discussed are not the only reasons that's so, by any means. Far more important is our common American culture. Although we are a continental nation, with a plurality of cultures, some of them regional, we are fortunate indeed that the values we have in common outweigh the shadings and nuances that divide us.

One of the most important of those values is our common commitment to the pursuit of happiness—which usually translates as the pursuit of money. Some critics of American culture, foreign and domestic, despise our "materialism," but the older I get the more wisdom I see in Samuel Johnson's observation that a man is seldom more innocently engaged than when he is making money. Surely it is no accident that the most successful multi-ethnic society on earth is the one that probably best exemplifies stodgy, money-making, bourgeois values: I am referring of course to Switzerland.

Once upon a time, Southerners—some of them, anyway—regarded materialism as a Yankee vice. Even if that was more than just propaganda, however, we have rejoined the Union in that respect. Back in the countercultural sixties, David Riesman observed that the only students he had at Harvard who wanted to make a lot of money were Catholics and Southerners. And survey research shows at least no difference in cupidity between Southerners and other Americans.

This common value, combined with considerable overlap of economic interest between settlers and natives, puts a damper on conflict. Let me close with one last story. Some time ago, the letters column of *The State* of Columbia witnessed a spirited exchange over whether Northern migration was good for South Carolina. A number of natives argued that it wasn't; even more migrants heatedly replied that it was. When the correspondence showed no signs of abating, *The State* closed the discussion with an editorial entitled "We Like Yankees."

The reason it gave—the *only* reason—was their connection with industrial development.

In Search of the Elusive Southerner

■ The woman was developing an orientation program for newcomers to the South, and she had come to see me because she'd heard that I'd written some books. "What does someone moving to the South need to know about the Southerner?" she asked. She waited expectantly.

I told her I couldn't possibly answer that question, that no one could say anything intelligent about "the Southerner." To begin with, I said, it matters where you're going in the South. Good advice for Tampa might not help at all in Austin, or Montgomery, or Hilton Head, or Bristol. W. J. Cash wrote a half-century ago that there are many Souths, and that may be even more true now. The South incorporates several different economies, many different landscapes. It has some of America's largest and fastest-growing metropolitan areas, with some of the country's wealthiest suburbs. It has some of the nation's poorest counties, which have lost population for decades. It has university towns and military bases, elegant resorts and tacky ones. Even among its rural communities and small market towns there are obvious differences: the mountains aren't the piedmont, and both of those are different from the coastal plain. The Southwest is something else again. And then there's Florida.

Consequently, I said, there's no single "Southerner" for newcomers to know about. A Kentucky mountaineer, an old-family Charlestonian, a Texas wildcatter, an Alabama tenant farmer, an Atlanta businesswoman—what could all these people have in common? Many non-Southerners don't recognize what a diverse place the South is. For that matter, neither do many Southerners.

Even if someone's ideas about Southerners are accurate for some parts of the region, there will be a great many exceptions. And some images aren't accurate anywhere, anymore, because they don't reflect how fast the South has changed in the last generation. Some Americans really do still think of the South as a romantic land of moonlight and magnolias; if they come here, they're in for some surprises. So are those whose stereotypes run more to pellagra and kukluxery. Neither image has much foundation in today's Southern reality, especially not in those parts where migrants are likely to be going.

I urged my visitor just to consider the changes of the last twenty-five years. Maybe by the 1960s it was evident that the South's future would be that of an urban, industrial region, but it hadn't really sunk in. Now nearly three out of four Southerners live in urban areas, and —this is important—most have never lived anywhere else. As recently as two generations ago, farmers were half of the South's labor force; now just one Southerner in twenty or so actually works on a farm, and most Southerners see farms only from the Interstate. A sure sign of the South's development is that our birthrate is no longer at an agricultural society's high level; in fact, it's now lower than the national average.

To be sure, some parts of our region have been largely bypassed by the South's development, and they have the same problems they've always had. But most of the South now has the problems of a developed economy: problems of growth, not stagnation.

Changes in race relations have been even more astonishing. When the Voting Rights Act was passed in 1965, who would have guessed that anyone would soon be arguing that race relations were better in the South than elsewhere? But now many Southerners, both black and white, are doing just that, and they're not just whistling Dixie. None of the many changes the Voting Rights Act produced is more striking than the fact that more blacks now hold public office in the South than in any other region. Mississippi has more black elected officials than any other state. Meanwhile, average black incomes in the South have surpassed those in the Midwest, and they increased during the 1980s while declining elsewhere in the United States. Perhaps the most persuasive testimony to the extent of these changes is that more blacks have moved to the South since the early 1970s than have left it, reversing a migration flow that in 1965 still seemed one of the unchangeable facts of American life.

These are only a few of the most obvious social and economic changes of the past few decades, I told my guest. Some changes were long-awaited, others were completely unexpected, but most of them run counter to longstanding stereotypes. Consequently, I suggested, newcomers would be well advised to come with open minds, and open eyes. They shouldn't think they know much about Southerners just because they've seen *Gone with the Wind,* "The Dukes of Hazzard," *Mississippi Burning,* or even "Designing Women." One thing few Southerners appreciate is being on the receiving end of uninformed stereotypes.

But, I said, we also don't appreciate hearing from newcomers that the South is no different from other parts of the country. That's because most Southerners believe that the South is *better.* Indeed, most seem to believe that they live in the best part of the South. Newcomers will find that there's a lot of regional pride in the South, and local pride, too. Country music's Tom T. Hall sings that "Country is . . . loving your town." But bragging on your town isn't just a country trait. Nobody does more of it than Atlantans.

The woman asked me why Southerners think the South is such a good place to live.

Well, I told her, look at *Southern Living*'s very first issue. It said that Southerners "live better" because they can "use and enjoy the South's open country, its mild climate, long growing season, and relatively uncrowded highways." The highways may be more crowded than they were in 1965, but there's still a lot of open country, and they can't take the climate away from us. Southerners still tell public opinion pollsters that the South offers superior living conditions.

And something else most Southerners say they like about the South is other Southerners. When you ask what we like about each other you get answers ranging from accents to values, but one thing in particular gets mentioned over and over again: Southerners describe each other as polite, courteous, friendly.

This business of Southern manners hasn't really been explored, I told her, but it's certainly something both natives and newcomers often notice and comment on. And I think they're right about that, I said. There are regional differences in manners—not good as opposed to bad, just Southern as opposed to twentieth-century American in general.

Most of us find that Southern manners make life pretty pleasant, day to day, but they can confuse someone who doesn't understand them. Surely everyone knows that "You all come see us" doesn't mean you should actually drop in, but maybe not everyone understands that when repairmen don't show up as promised it may be because they think it would be impolite to tell you they're too busy to come.

In general, it seems to me that Southerners often go to some lengths to avoid disappointing people, and even more to avoid direct criticism and disagreement—that is, unless they *mean* to be offensive. (Excuse me for saying this, I told my visitor, but it follows that Southerners don't like to be criticized, even when critics mean to be helpful. Maybe newcomers ought to know that.)

"Why can't Southerners just say what they mean?" she asked.

But we do, I said. At least we usually understand each other. It's just a matter of how you say what you mean. You can see the same sort of round-about approach in Southern humor. It isn't that Southerners can't be amused by Woody Allen or Rodney Dangerfield or Eddie Murphy, but there's also a peculiarly Southern form of humor that doesn't rely on one-liners, or on wit. It's a matter of storytelling, where the humor is in the details and the style. As Jerry Clower put it once: "I don't tell funny stories. I tell stories funny." This takes some getting used to, and non-Southerners sometimes miss the point. Same with manners.

Some other Americans who like to think of themselves as straightforward, no-nonsense, and businesslike have been known to complain about Southern indirection. They wonder why Southerners can't just get to the point. But Southerners are likely to think that there are usually several points, one of them being to put the person you're dealing with at ease. So even in the business world there may be a tendency—just a tendency, that's all—to the sort of preliminary small talk and inquiries about family and so forth that are more common in Oriental bazaars than in the commerce of the Northeast.

The absence of those pleasantries can be misinterpreted, too, as when some Southerners complain about newcomers who order people around "like they owned them." Small talk can serve to make the point that just because you're paying someone doesn't mean you think you're better than he is. In this and other ways, Southern manners are egalitarian. I mentioned W. J. Cash again. He wrote in

The Mind of the South about the old-time Southern mill owner, mixing with the workers, slapping backs and knowing names, respecting their dignity as human beings and Christian souls.

And speaking of religion, I said, something else that strikes many newcomers to the South is the part that churches play in the South's public and social life. They ought to be prepared for that. Certainly you'd be making a mistake to come to most parts of the South if you mind living in a place where most people are Baptists and Methodists. It would be like moving to Utah if you don't like Mormons, or Massachusetts if Roman Catholics get on your nerves.

You don't have to be Baptist or Methodist or even Protestant yourself; these days you can be whatever you want. But it helps if you're *something*. Belonging to a church, and being more or less active in it, is a taken-for-granted part of middle-class life in the South, in a way that it's not in many other parts of the country. Nearly everybody, rich or poor, urban or rural or suburban, black or white, has a church to go to. Even those Southerners who don't go to church at least know which one they're not going to.

Religion is tied in with race and with social class, in complicated ways that Southerners understand but don't talk about much. In that, it's like hunting and fishing. Different folks do these things in different ways, according to how they were raised and what their circumstances encourage. Trout fishing is something like Episcopalianism—sort of the Anglicanism of angling, if you'll excuse the expression. Going after catfish with a cane pole and chicken innards may correspond to Holiness. Somewhere in between lies the market for bass boats and electronic fish-finders. Just so, some Southerners hunt quail, while others hunt squirrel, and coon hunters are different from the Ducks Unlimited crowd. But they're all hunters, and there are a lot of them. You don't have to hunt and fish if you live in the South, but if you don't like hunters, you might be happier somewhere else. Same with religion.

And that reminds me of sports, I said. There is something almost religious about Southerners' attachment to their teams. In small towns, going to the high school game on Friday night is almost on a par with going to church on Sunday. It's the same with college sports. Where else would a magazine like *Southern Living* pick an all-star football team each fall? But it makes sense: a great deal of Southern social life does seem to revolve around tailgate parties. And those

Southern cities that are big enough to have professional teams—well, they may not win very often, but they've got plenty of loyal fans.

Newcomers might want to pick a team and follow it. It doesn't greatly matter which one—it's like religion that way, too. Again, you don't have to be interested in sports to live in the South, but life may be easier if you are, or can pretend to be. Even in the business world, much of that small talk I mentioned earlier has to do with sports. A newcomer who doesn't know the New Orleans Saints from the Macon Whoopees might find it a good investment (if nothing else) to learn.

Of course much of this is just speculation, I told my visitor, and even if all of it's true it may be changing. One constant in the South has been change. That's a paradox, isn't it?

I looked at the time, and realized I had another appointment. So, I said, I'm sorry, but there's really no way to say what people moving to the South ought to know about Southerners. You can't begin to generalize, and even if you could I wouldn't do it. Southerners don't much like folks generalizing about them.

Bibliography

■ Baldwin, Joseph G. *The Flush Times of Alabama and Mississippi: A Series of Sketches*. New York: D. Appleton, 1853.

Bertelson, David. *The Lazy South*. New York: Oxford University Press, 1967.

Black, Merle. "Is North Carolina Really the 'Best' American State?" In *Politics and Policy in North Carolina,* edited by Thad L. Beyle and Merle Black. New York: MSS Publications, 1975.

———. "The Modification of a Major Cultural Belief: Declining Support for 'Strict Segregation' Among White Southerners, 1961–1972." *Journal of the North Carolina Political Science Association* 1 (Summer 1979): 4–21.

Blu, Karen I. *The Lumbee Problem: The Making of an American Indian People*. New York: Cambridge University Press, 1980.

Boles, John B. "Laments Are Premature for Southern Regionalism." *Humanities in the South,* no. 58 (Fall 1983): 2 ff.

Botsch, Robert. *We Shall Not Overcome: Populism and Southern Blue-Collar Workers.* Chapel Hill: University of North Carolina Press, 1980.

Brearley, H. C. "Are Southerners Really Lazy?" *American Scholar* 18 (Winter 1949): 68–75.

Caldwell, Erskine, and Margaret Bourke-White. *You Have Seen Their Faces*. New York: Modern Age Books, 1937.

Capote, Truman. *Answered Prayers: The Unfinished Novel*. New York: Random House, 1987.

Cash, W. J. *The Mind of the South*. New York: Alfred A. Knopf, 1941.

Duberman, Martin B. *Black Mountain: An Exploration in Community.* New York: Dutton, 1972.

Ehrenreich, Barbara. *The Hearts of Men: American Dreams and the Flight from Commitment*. Garden City, N.Y.: Anchor Press/Doubleday, 1983.

Freymeyer, Robert H. "Republicans Flow South." *American Demographics* 4 (June 1982): 35–37.

Gans, Herbert. *The Urban Villagers: Group and Class in the Life of Italian-Americans.* New York: Free Press of Glencoe, 1962.

Garreau, Joel. *The Nine Nations of North America.* Boston: Houghton Mifflin, 1981.

Garrett, George. "Southern Literature Here and Now." In Fifteen Southerners, *Why the South Will Survive.* Athens: University of Georgia Press, 1981.

Gaston, Paul. *The New South Creed: A Study in Southern Mythmaking.* Baton Rouge: Louisiana State University Press, 1976.

Gee, Wilson. "The 'Drag' of Talent out of the South." *Social Forces* 15 (1937): 343–46.

Glenn, Norval D. "Massification versus Differentiation: Some Trend Data from National Surveys." *Social Forces* 46 (1967): 172–80.

Glenn, Norval D., and J. L. Simmons. "Are Regional Cultural Differences Diminishing?" *Public Opinion Quarterly* 31 (1967): 176–93.

Glenn, Norval D., and Charles Weaver. "Regional Differences in Attitudes toward Work." *Texas Business Review* 56 (November–December 1982): 263–66.

Gould, Peter, and Rodney White. *Mental Maps.* Harmondsworth: Penguin Books, 1974.

Grady, Henry W. *The New South.* New York: Robert Bonner's Sons, 1890.

Grasmick, Harold. "Social Change and the Wallace Movement in the South." Ph.D. dissertation, University of North Carolina at Chapel Hill, 1973.

Greeley, Andrew M. *Ethnicity in the United States: A Preliminary Reconnaissance.* New York: John Wiley & Sons, 1974.

Havard, William C., Jr. "Southern Politics: Old and New Style." In *The American South: Portrait of a Culture,* edited by Louis D. Rubin, Jr. Baton Rouge: Louisiana State University Press, 1980.

Hobson, Fred. *Tell about the South: The Southern Rage to Explain.* Baton Rouge: Louisiana State University Press, 1983.

Holloway, Harry, and Ted Robinson. "The Abiding South: White Attitudes and Regionalism Reexamined." In *Perspectives on the American South,* edited by Merle Black and John Shelton Reed, vol. 1. New York: Gordon and Breach, 1981.

Horton, Hamilton C., Jr. "The Enduring Soil." In Fifteen Southerners, *Why the South Will Survive.* Athens: University of Georgia Press, 1981.

Hunnicutt, Benjamin K. *See* Reed, John Shelton, and Benjamin K. Hunnicutt.

Hurlbert, Jeanne S. "The Southern Region: A Test of the Hypothesis of Cultural Distinctiveness." *Sociological Quarterly* 30 (1989): 245–66.

Kallen, Horace M. "Democracy versus the Melting Pot: A Study of American Nationality." *The Nation* 100 (February 18–25, 1915): 190–94, 217–20.

Killian, Lewis. *White Southerners*. Rev. ed. Amherst: University of Massachusetts Press, 1985.

Kim, Choong Soon. *An Asian Anthropologist in the South: Field Experiences with Blacks, Indians, Whites*. Knoxville: University of Tennessee Press, 1977.

Klapp, Orrin E. *Heroes, Villains, and Fools: The Changing American Character.* Englewood Cliffs, N.J.: Prentice-Hall, 1962.

Lazarsfeld, Paul F., and Wagner Thielens, Jr. *The Academic Mind: Social Scientists in a Time of Crisis*. Glencoe, Ill.: The Free Press, 1958.

Loewen, James W., and Charles Sallis, eds. *Mississippi: Conflict and Change*. New York: Pantheon Books, 1974.

Longino, Charles, and Jeanne Biggar. "The Impact of Retirement Migration on the South." *The Gerontologist* 21 (1981): 283–90.

Luebke, Paul. *Tar Heel Politics: Myths and Realities*. Chapel Hill: University of North Carolina Press, 1990.

McAllister, Ward. *Society As I Have Found It*. New York: Cassell Publishing Co., 1890.

McWhiney, Grady. *Cracker Culture: Celtic Ways in the Old South*. University, Alabama: University of Alabama Press, 1988.

Malone, Bill C. *Southern Music: American Music*. Lexington: University Press of Kentucky, 1979.

Marsden, Peter V., John Shelton Reed, Michael D. Kennedy, and Kandi M. Stinson. "American Regional Cultures and Differences in Leisure Time Activities." *Social Forces* 60 (1982): 1023–49.

Marx, Karl, and Frederick Engels. *The Civil War in the United States*. New York: Citadel Press, 1961.

Matthews, Donald R., and James W. Prothro. *Negroes and the New Southern Politics*. New York: Harcourt, Brace & World, 1966.

Mencken, H. L. "The Sahara of the Bozart" (1917). In *Prejudices, Second Series*. New York: Alfred A. Knopf, 1920.

Mitchell, Broadus. "Fleshpots in the South." In *The Industrial Revolution in the South,* by Broadus Mitchell and George Sinclair Mitchell. Baltimore: Johns Hopkins Press, 1930.

Morris, Willie. *North toward Home*. Boston: Houghton Mifflin, 1967.

Murray, Albert. *South to a Very Old Place*. New York: McGraw-Hill, 1971.

O'Brien, Michael. *The Idea of the American South, 1920–1941.* Baltimore: Johns Hopkins University Press, 1979.

Odum, Howard W. *Southern Regions of the United States.* Chapel Hill: University of North Carolina Press, 1936.

Olmsted, Frederick Law. *The Cotton Kingdom: A Traveller's Observations on Cotton and Slavery in the American Slave States . . .* 2d ed. New York: Mason Brothers, 1862.

Percy, Walker. "Questions They Never Asked Me." *Esquire,* December 1977, 170 ff.

Phillips, Ulrich Bonnell. "The Central Theme of Southern History." *American Historical Review* 34 (October 1928): 30–43.

———. *Life and Labor in the Old South.* New York: Grosset & Dunlap, 1929.

Pinckney, Josephine. "Bulwarks against Change." In *Culture in the South,* edited by W. T. Couch. Chapel Hill: University of North Carolina Press, 1934.

Potter, David M. *The South and the Concurrent Majority.* Baton Rouge: Louisiana State University Press, 1972.

Powledge, Fred. *Journeys through the South: A Rediscovery.* New York: Vanguard Press, 1979.

Reed, John Shelton. "The Cardinal Test of a Southerner?" In *One South: An Ethnic Approach to Regional Culture.* Louisiana State University Press, 1982.

———. *The Enduring South: Subcultural Persistence in Mass Society.* Rev. ed. Chapel Hill: University of North Carolina Press, 1986.

———. "In with the In-Crowd: Talkin' Trash, Spendin' Cash." In *Whistling Dixie: Dispatches from the South.* Columbia: University of Missouri Press, 1990.

———. *One South: An Ethnic Approach to Regional Culture.* Louisiana State University Press, 1982.

———. *Southerners: The Social Psychology of Sectionalism.* Chapel Hill: University of North Carolina Press, 1983.

———. *Southern Folk, Plain & Fancy: Native White Social Types.* Athens: University of Georgia Press, 1986.

Reed, John Shelton, and Merle Black. "Blacks and Southerners." In *One South: An Ethnic Approach to Regional Culture.* Louisiana State University Press, 1982.

———. "How Southerners Gave Up Jim Crow." *New Perspectives,* Fall 1985, 15–19.

Reed, John Shelton, and Benjamin K. Hunnicutt. "Leisure." In *The Ency-*

clopedia of Southern Culture, edited by Charles Reagan Wilson and William Ferris. Chapel Hill: University of North Carolina Press, 1989.

Reed, Roy. "Revisiting the Southern Mind." *New York Times Magazine,* December 5, 1976, 42–43, 99–109.

Rindfuss, Ronald R. "Changing Patterns of Fertility in the South: A Social-Demographic Examination." *Social Forces* 57 (1978): 621–35.

Rindfuss, Ronald R., John Shelton Reed, and Craig St. John. "A Fertility Reaction to a Historical Event: White Southern Birthrates and the 1954 Desegregation Decision." *Science* 201 (July 14, 1978): 178–180.

Royce, Josiah. *Race Questions, Provincialism, and Other American Problems.* New York: The Macmillan Co., 1908.

Rubin, Louis D., Jr. "The American South: The Continuity of Self-Definition." In *The American South: Portrait of a Culture,* edited by Louis D. Rubin, Jr. Baton Rouge: Louisiana State University Press, 1980.

Smiley, David L. "The Quest for the Central Theme in Southern History." Paper read at Conference of the Southern Historical Association, 1962.

Snow, Ronald W. "Recent Migrants to Mississippi." In *Perspectives on the American South,* edited by Merle Black and John Shelton Reed, vol. 1. New York: Gordon and Breach, 1981.

Stadiem, William. *A Class by Themselves: The Untold Story of the Great Southern Families.* New York: Crown Publishers, 1980.

Still, William. *Still's Underground Rail Road Records: With a Life of the Author . . .* Rev. ed. Philadelphia: William Still, 1883.

Talese, Gay. *Thy Neighbor's Wife.* Garden City, N.Y.: Doubleday, 1980.

Target Group Index, Spring, 1976. 52 vols. New York: Axiom Market Research Bureau, 1976.

Thompson, Edgar T. *Plantation Societies, Race Relations, and the South: The Regimentation of Populations. Selected Papers of Edgar T. Thompson.* Durham: Duke University Press, 1975.

Thompson, Hunter S. *Hell's Angels: A Strange and Terrible Saga.* New York: Ballantine Books, 1967.

Tindall, George Brown. *The Ethnic Southerners.* Baton Rouge: Louisiana State University Press, 1976.

Trudgill, Peter. *Sociolinguistics: An Introduction to Language and Society.* Harmondsworth: Penguin Books, 1983.

Twelve Southerners. *I'll Take My Stand: The South and the Agrarian Tradition.* New York: Harper & Brothers, 1930.

Wood, Gail. "The Images of Southern Males and Females." Senior honors paper, University of North Carolina at Chapel Hill, 1973.

Woodward, C. Vann. *The Burden of Southern History.* Rev. ed. Baton Rouge: Louisiana State University Press, 1968.

———. "W. J. Cash Reconsidered." *New York Review of Books* 13 (December 4, 1969): 28–34.

Wyatt-Brown, Bertram. *Southern Honor: Ethics and Behavior in the Old South.* New York: Oxford University Press, 1982.

Yoder, Edwin M., Jr. *The Night of the Old South Ball, and Other Essays and Fables.* Oxford, Mississippi: Yoknapatawpha Press, 1984.

Zelinsky, Wilbur. "Selfward Bound? Personal Preference Patterns and the Changing Map of American Society." *Economic Geography* 50 (April 1974): 144–79.